C H I N A 2 0 2 0

FINANCING HEALTH CARE

CHINA 2020 SERIES

China 2020:
Development Challenges in the New Century

Clear Water, Blue Skies:
China's Environment in the New Century

At China's Table:
Food Security Options

Financing Health Care:
Issues and Options for China

Sharing Rising Incomes:
Disparities in China

Old Age Security:
Pension Reform in China

China Engaged:
Integration with the Global Economy

THE WORLD BANK
WASHINGTON D.C.

FINANCING HEALTH CARE

ISSUES AND OPTIONS FOR CHINA

THE WORLD BANK
WASHINGTON D.C.

First printing September 1997

Cover photograph by Erica Lansner/Black Star.

Cover insets (from left to right) by Vince Streano/Aristock, Inc.; Claus Meyer/Black Star; Serge Attal/Gamma Liaison; Dennis Cox/China Stock; Joe Carini/Pacific Stock; Dennis Cox/China Stock.

ISBN: 0-8213-4048-4

Contents

This report uses *Hong Kong* when referring to the Hong Kong Special Administrative Region, People's Republic of China.

Acknowledgments

This study was prepared by the World Bank at the request of the government of the People's Republic of China, which is considering policies to improve the equity and efficiency of health services. It is the third World Bank policy study of China's health sector. The series began with a general study entitled *China: The Health Sector* (1984), followed by *China: Long-Term Issues and Options in the Health Transition* (1992a). This report looks in greater detail at the impact of China's move toward a market economy on the financing and organization of health care and recommends a number of steps China can take to reform health care financing.

Research for this report was carried out jointly by the Chinese government and World Bank staff. Work in China proceeded under the direction of Liu Peilong, director of the Foreign Loan Office of the Ministry of Health. Two World Bank task managers in the Human Development Department managed the project: William

P. McGreevey, from June 1994 to February 1996, and Helen Saxenian, from March 1996 to August 1997. The report was written by Helen Saxenian, together with William Hsiao, Dean T. Jamison, William P. McGreevey, and Winnie Yip. Work at the World Bank was carried out under the direction initially of Vinay Bhargava, chief, Human Resource Division, China and Mongolia Department, followed by Joseph Goldberg, chief, Rural and Social Development Division. Jagadish Upadhyay, health group manager, and Janet Hohnen, public health specialist, managed the work on behalf of the division. Nicholas Hope, director, China and Mongolia Department, guided the preparation work. Richard Newfarmer, lead economist in the department, and Michael Walton, chief economist, East Asia and Pacific Region, helped set the overall context for the report. William P. McGreevey and Helen Saxenian worked under the general direction of Richard Feachem, senior adviser, Human Development Department. World Bank staff in Beijing, including Pieter Bottelier, resident mission director, Ramgopal Agarwala, Kathy Ogawa, and Zhao Hongwen, also provided assistance.

China's minister of health, Chen Minzhang, guided the early report preparation in October 1994. Ministry of Health staff, including Liu Xinming, deputy director, Planning and Finance Department, Cai Renhua, director, Legal Affairs Department, Liu Yingli and Liu Junguo of the Foreign Loan Office, and Fei Zhao Hui, now of the Ministry of Finance, provided considerable help during field visits. Mme. Sun, Ministry of Finance, and Ying Li, State Council, reviewed many phases of the study's work and provided advice throughout.

The study draws extensively on fifteen background papers prepared between November 1994 and September 1995 by leading specialists in China's health economics and finance. A list of the papers and their authors appears before the bibliography.

A related study of China's national health accounts was partially funded by the World Bank and a special grant from the Canadian government. Contributors to that study include Peter Berman of Harvard University, Gilles Fortin of the Canadian Institute for Health Information, Vernon Hicks of Health Economics Consulting Services in Halifax, Nova Scotia, and J. Brad Schwartz.

An advisory group has been helpful throughout in defining the principal issues and approaches in this study. This group consists of William Hsiao; K. T. Li, professor of health economics at Harvard University; Hu Shanlian of Shanghai Medical University and deputy director, Health Economics Institute, Beijing; Dean T. Jamison, director, Center for Pacific Rim Studies, University of California at Los Angeles; and Wei Ying, Beijing Medical University and director, Health Economics Institute.

Peer reviewers in the World Bank include Willy de Geyndt, Charles Griffin, Jeffrey Hammer, Emmanuel Jimenez, and Nicholas Prescott. Members of the World Bank's Health Group in the Human Development Department provided valuable comments, including Howard Barnum, Denis Broun, Philip Musgrove, Mary Young, and George Schieber. Richard Bumgarner, Yuanli Liu, and Richard Peto also provided valuable input.

Bank staff visited the provinces of Shanxi and Jiangsu in October 1994, and Sichuan, Hebei, Jiangxi, and Guizhou in April 1995. Bank staff members and consultants who joined these missions and contributed to the report include John S. Akin, Harry E. Cross, Jeffrey Hammer, Winnie Yip, and Zhou Ji An. Helene Genest and Paul Hutchinson provided valuable assistance to the mission work. An initial draft of this report was discussed with the Chinese government in October 1995. Follow-up work on national health accounts was done during a December 1995 mission. The full report was discussed with Chinese government officials in a workshop on 22–24 April 1996 and this version incorporates their comments, as well as further written comments received through July 1997.

An earlier version of this report was edited by Madelyn Ross. Mylene Domingo prepared the manuscript with the help of Susan Sebastian, Akosua Hudgens, Yvette Atkins, and Euna Osbourne. Jillian Cohen, Ellen Lukens, and other Health Group staff helped in checking sources and manuscript review.

The book was edited by Alison Strong, designed by Kim Bieler, and laid out by Glenn McGrath and Damon Iacovelli of the American Writing Division of Communications Development Incorporated.

Overview

Before 1949 China's population was among the least healthy in the world. Its poor health was both a consequence and a cause of the nation's poor economic performance. China's investments in improving health since then have directly enhanced well-being, particularly among the poor, while also contributing to rapid economic growth. The country's dramatic success in improving health conditions—as reflected in life expectancy's rise from less than forty years in 1950 to sixty-nine years in 1982—was accompanied by two related but less frequently noted achievements:

• By 1975 insurance coverage (provided by the government and state enterprises) and the rural cooperative medical system had reached close to 90 percent of the population—almost all the urban population and 85 percent of the rural. Although this coverage was not without major problems, it did provide China's citizens with some access to cost-effective preventive and curative health services and some sharing of the risks of medically caused financial misfortune.

- The system for finance and delivery of health services contained costs. In 1981 health care costs were just over 3 percent of GDP, despite the remarkable gains in health and in insurance coverage.

Beginning in 1978, the Chinese government introduced radical economic policy shifts that moved China away from a centrally planned economy and toward a competitive market system. This change in economic policies was accompanied by a devolution of power to provincial governments. Many of the changes have had profound repercussions for the health system. In rural areas the transition from agricultural collectives to the household responsibility system weakened the financial base of the cooperative medical system. In the health sector the government has encouraged programs and facilities to rely on user fees to support their operations, but continues to administer many prices, setting most below cost, and to control staffing in public facilities.

Problems in health sector performance

China's many achievements in health over the past several decades have been recognized internationally—its improvements in health status, its broadening of physical access to basic health services, and its support of important public health measures. But its health sector faces deep problems today, as measured by financial access to health care, by efficiency, and by total cost. The trend in child mortality, an important indicator of health outcomes, also appears to be a cause for concern. Some of these problems are common to many countries. Others relate to the government's failure to reformulate health finance and to redefine its role in health. China needs to act now to correct these problems, before they become more deeply rooted. The action needs to be at a high level and interministerial. Health is a sector that cannot simply be left to market forces.

Much progress and some problems in health status

China's overall health status, as measured by life expectancy and infant, child, and maternal mortality, is excellent compared with that of other countries at similar income levels. But recent trends in child mortality are less clear. Estimates derived from fertility and population census data suggest that after falling steadily for forty years, China's under-five mortality rate appears to have leveled off in the mid-1980s at about 44 per 1,000 live births (figure 1). But death registration data from China's Ministry of Health indicate that under-five mortality declined in the 1990s—from 61 per 1,000 live births in 1991 to 51 per 1,000 in 1995. The different results from these two methods for estimating the under-five mortality rate suggest that more detailed analysis is required to understand the true trends.

Growing disparity in financial access to health care

People in China have relatively good physical access to basic health care services. High population density and a well-developed health infrastructure mean that geographical barriers are modest for all but a significant minority living in mountainous or remote rural areas. The cost of routine, basic outpatient health services is low enough that most nonpoor Chinese households can pay for them out of current income or savings. Beyond that, however, financial access to health care in China is inequitable, with especially deep divisions between the urban and rural population.

For China as a whole, health spending per capita (public and private) was estimated at 110 yuan, or $13.50, in 1993. (Health spending in purchasing power parity terms would be 4.8 times higher because of inter-

FIGURE 1

Steady decline, then a leveling off in China's under-five mortality rate

Deaths per 1,000 live births, 1960–91

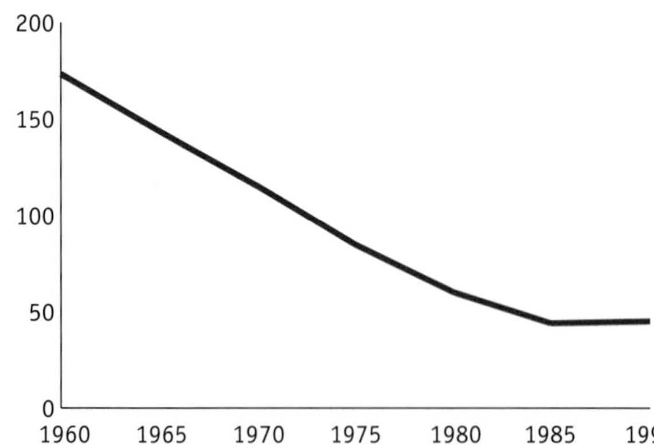

Note: Rates are estimates based on survey data.
Source: Hill and Maeda 1997.

FIGURE 2

Inequitable financial access to health care for rural Chinese

Population by insurance status, 1993

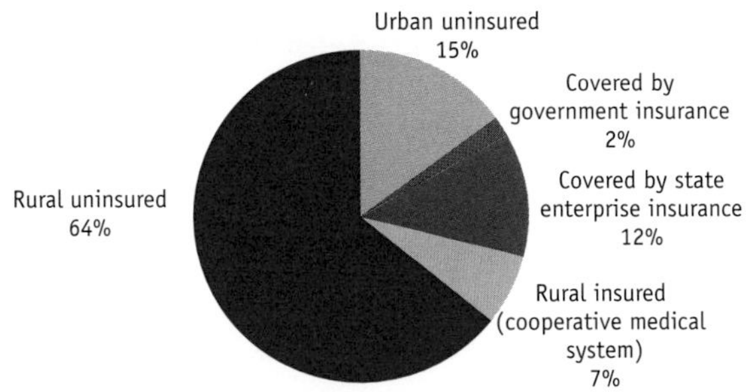

Health expenditure by source of finance, 1993

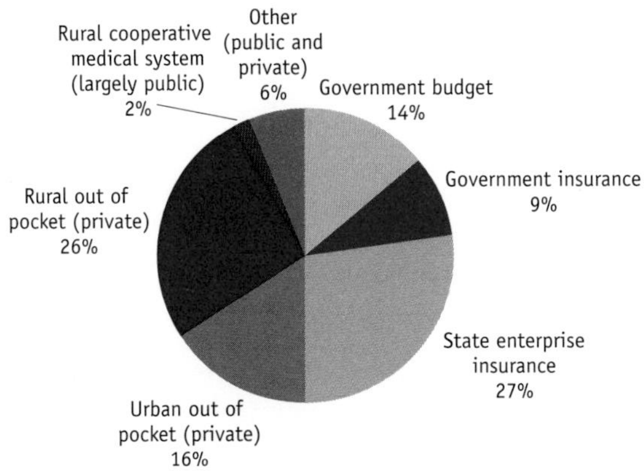

Source: Wei 1995.

national price differences.) But average health spending in urban areas, at 235 yuan per capita, was almost four times the average of 60 yuan in rural areas. And the poorest quarter of the rural population accounted for only about 5 percent of all health spending in 1993. Only 10 percent of the rural population is insured, compared with 50 percent of the urban population. While the two urban health insurance systems—the government and state enterprise insurance systems—cover only 15 percent of China's population, they absorb two-thirds of public spending on health and 36 percent of all health spending (figure 2).

Coverage under the cooperative medical system in rural areas has declined rapidly since the late 1970s, largely because of the introduction of the rural production contract responsibility system. The shift away from a communal system deprived the rural cooperative medical system of its sources of community-based financing. As communes gradually disappeared, so did the cooperative medical system. Only about 10 percent of the rural population is now covered by some form of community-financed health care, down from a peak of about 85 percent in 1975. (There is much variation in coverage across provinces, however, because of differences in interpretation of national policy.) As a result, some 700 million rural Chinese must pay out of pocket for virtually all health services. Without insurance, medical expenses can lead to deferral of care, untreated illness, financial catastrophe, and poverty.

Increasing inefficiency

A long-standing problem in China is duplication of facilities and the excess capacity in some vertical national health programs. In urban areas there is overlap among Ministry of Health, state enterprise, and traditional Chinese medicine facilities. In rural areas there are growing duplication and overlap of services between maternal and child health centers, family planning services, township health centers, and epidemic prevention stations. This overlap results in inefficiency and waste.

Public spending on health is skewed toward hospitals, even as priority public health programs are increasingly underfunded. With fiscal decentralization, the poorest counties have become least able to finance public health programs. As a result of funding difficulties, some public health workers have been diverted from important public health work, such as immunization and disease surveillance, to activities for which they can more easily charge fees, such as routine testing of food and water in urban areas. And the Epidemic Prevention Service is now charging for immunizations and tuberculosis treatment in many parts of the country. This practice has reduced coverage and, in tuberculosis treatment, led to medically inappropriate but profitable patterns of care.

Prices of most health services and many inputs to the health sector are fixed well below cost under guidelines issued by the Price Commission. To cross-subsidize underpriced products and services and to generate profits, health care providers inappropriately promote profitable items—especially pharmaceuticals and high-technology diagnostic tests. This leads to misallocation of spending, medically inappropriate services, and upward pressure on health spending in both rural and urban areas. Given the incentive structure, it is not surprising to find that pharmaceuticals account for a remarkably high share of health spending—52 percent in 1993.

Rising cost of health care

Total health spending per capita grew 8 percent a year in real terms from 1978 to 1986, accelerating to 11 percent a year from 1986 to 1993. Over the same period real GDP per capita grew 7.7 percent a year. Health spending now accounts for about 3.8 percent of GDP. Spending will continue to grow in real terms as China's income grows, and this growth is likely to be accelerated by price distortions in the health sector and heavy reliance on fee-for-service provider payment—particularly fee-for-service under third-party insurance systems such as the government and state enterprise systems. Spending growth in these two insurance systems is simply not sustainable.

The aging of the population will also increase health spending, because the elderly have higher health costs than the young. People aged sixty-five and over now make up 6 percent of China's population, and their share will reach 11 percent by 2020. While the aging of the population is inevitable, government policies can influence how efficiently the health care system addresses the needs of the elderly. And introducing effective health promotion and disease prevention programs now—particularly to reduce tobacco use—could do much to improve the health outlook for China's elderly.

Another part of the cost of health finance and delivery is the economic distortions that result from China's urban insurance systems. Because health coverage is tied to the employer—the government or a state enterprise—workers cannot retain their social benefits if they move from one job to another. Reforms are therefore needed so that workers can change jobs without jeopardizing their health (and pension) benefits.

Recommendations

Despite China's remarkable early and continuing successes in the health sector, issues of access, efficiency, and cost containment point to problems in the health sector's performance. In health indicators, the trend in under-five mortality appears to be a cause for concern. The Chinese government has reached a consensus that these important concerns must be addressed by strong policy initiatives. In December 1996 the State Council and the Central Committee of the Communist Party held a national health conference to discuss and examine major policy issues in health and later issued "Decisions on the Health Reform and Development." The rest of this overview presents recommendations for dealing with the issues China faces in its health sector and then discusses the implications of those recommendations for public expenditures.

Strengthen public health programs

Since the founding of the People's Republic in 1949, China has complemented the development of local health services with a series of strong national programs for high-priority public health activities, including disease surveillance, mass immunization, health education, and environmental monitoring and improvement. The government also supported the treatment of infectious diseases. Since most public health programs provide services that yield large social benefits, but for which individuals are unwilling to pay the full cost, financing these programs was an appropriate and critical role for the government.

Three related problems increasingly limit the effectiveness, scope, and coverage of China's national public health programs. First, budgetary pressures constrain the operation and efficiency of programs. The resource requirements are modest relative to total health spending. But almost all spending on public health is by provincial and local governments, so the poorest areas—which also have the worst public health problems—have the least capacity to finance these programs. Second, the policy emphasis on cost recovery has led to the introduction of user fees for some public health services (such as immunization), limiting demand for them, particularly among the poor. Third, the general movement toward fee-for-service payment has diverted an important part of the work of public health workers to activities for which fees can most easily be charged, rather than those with the highest priority for public health.

China needs to return to a policy of vigorous finance and support for public health, recognizing that these services must be financed by the government if they are to be provided at socially optimal levels. Particular attention needs to be given to reaching the unregistered urban population. This report recommends that the Epidemic Prevention Service's budget, which was 1.3 billion yuan in 1993, be increased to at least 6.5 billion yuan by 2001 and that the agency be prohibited from charging user fees for most of its services. Other agencies carrying out priority public health activities also need additional support. The government must also ensure that public health programs are implemented efficiently and that China's highly effective disease surveillance system is maintained and adapted to the changing pattern of disease.

A critical public health problem is the massive burden of costly illness and premature death from tobacco-related diseases. Nearly 1 million Chinese die each year of smoking-related diseases, and tobacco-related deaths are projected to increase to more than 2 million a year by 2020. Thus another recommended public health action is a substantial increase in the tobacco tax, accompanied by other measures to reduce smoking. Raising the tax would help reduce smoking-related illness and deaths and, if the incremental revenue were applied to public health (as in Australia), could help ensure adequate financing for public health programs.

Ensure essential health services for the poor

The second priority for government health spending should be to ensure that the country's neediest citizens have access to priority health services. The poor are more likely to suffer from ill health, and their health problems can keep them in poverty. Almost all of China's absolute poor live in rural areas. Among the poorest quarter of the rural population, infant mortality is 3.5 times greater than the rate among city dwellers. But the urban poor, especially unregistered migrants, also face high health risks, and they too need to be reached more effectively.

There is strong justification on poverty assistance or equity grounds for government subsidies aimed at improving the access of the poor to important health services. The Chinese government now provides minimal subsidies for this purpose. Resources need to be redirected or expanded to ensure key health services for the poor. And because public resources are scarce, subsidies need to be carefully targeted. There are several ways to target the poor:

- Geographical targeting to areas where the poor are concentrated. Subsidized services could be targeted to poor villages in China's 592 officially designated poor counties, for example, where the population totals about 75 million.
- Individual or household targeting, by identifying the poor and certifying their eligibility for subsidized services. (China may be the only developing country where this targeting method is feasible because of good government records, but this kind of targeting entails heavy administrative costs.)
- Program targeting to health services that particularly benefit the poor, in both rural and urban areas, such as deworming and management of acute infections in children. (This effort would require subsidies in addition to those for the public health programs discussed in the previous section.)

This report recommends phasing in a blend of geographical targeting (probably most practical at the village level) and program targeting for a few services that particularly benefit the poor. It also recommends monitoring these approaches to guide policy improvements.

Reform prices and provider payment mechanisms

Price distortions and irrational allocation of health resources have diminished the quality and effectiveness of China's health services. The government sets prices for most medical services well below cost. To allow health care providers to offset the resulting losses on basic services, the government has permitted them to charge high prices for drugs and high-technology diagnostic tests. The result is a distorted pattern of services, with overprovision of some, such as computerized tomography (CT) scans and ultrasounds, and underprovision of others, especially those with a high labor content.

Fee-for-service provider payment gives providers strong incentives to overprescribe drugs of all kinds, especially expensive drugs, in order to bring in additional income. It also encourages the overprovision of services. For example, outpatients are often treated with intravenous drip solutions of glucose, vitamins, antibiotics, and other drugs—treatment that in too many cases does not constitute justified medical practice. Spending on drugs accounts for more than half of all health spending in China, compared with 5–20 percent in OECD countries and 15–40 percent in most developing countries.

Price and provider payment reform is essential to help contain costs, reduce waste, and improve efficiency. Price reform should bring administered prices in the health sector in line with marginal costs. A major study of pricing is needed to lay the basis for this reform. To be politically acceptable, price and provider payment reform would probably need to be implemented gradually over several years. Price reform in the health sector is integral to solving the economywide problem of price distortions.

China also needs to move away from its dependence on fee-for-service provider reimbursement. Unconstrained fee-for-service reimbursement promotes excessive use of services, because consumers rely on providers to recommend the services they need, and providers have a financial incentive to increase the volume of services. Experience in other countries shows that case-based reimbursement and partial and full capitation payment methods can help contain costs and improve quality. Alternatively, fee-for-service payment methods can be used under a global budget constraint, or a mix of provider payment approaches can be used to improve incentive structures.

China's options for provider payment reform hinge on the types of risk-pooling arrangements developed in rural and urban areas. One option would be to move—initially under urban insurance and community financing schemes—from fee-for-service toward more aggregated products, an approach tested in Zhenjiang, and finally to prepayment for a complete package of services. Under a prepayment system the provider would assume more risk and would have a decreasing incentive to overprovide services.

Control investments and improve regional planning

Governments can play an important part in containing health costs through oversight and control over major human and capital investments in the health system. Although China is moving away from a centrally planned economy, the government needs to retain some oversight over new investments in hospital beds (especially at the tertiary level) and expensive medical equipment and over the mix and numbers of health personnel. Experience in other countries shows that, once created, excess supply in any of these areas is politically difficult to correct. This excess capacity drives up spending through supplier-induced demand.

The government could complement such supply-side controls with efforts to improve regional planning in health. China's health system is plagued by poorly coordinated vertical delivery systems in both rural and urban areas. Regional planning efforts should involve all relevant actors, including medical schools, the Ministry of Health, traditional Chinese medicine facilities, and the government and state enterprise insurance systems. As urban insurance centers expand, they would also become major stakeholders in regional planning efforts. China can build on experience from planning exercises already under way in Baoji in Shaanxi Province, Jiujiang in Jiangxi Province, and Jinhua in Zhejiang Province.

Promote efficient risk pooling in rural and urban areas

In the poorest countries of the world more than half of all health spending comes from private sources, mainly out-of-pocket expenditures, and the poor pay for most of their services. As national income rises, the share of out-of-pocket health spending gradually falls as mechanisms develop for pooling the risk of catastrophic health expenses. Government support to encourage the development of risk-pooling mechanisms can make health services more accessible and efficient. Risk-pooling mechanisms can be financed by general tax revenue, social health insurance (mandated payroll taxes), private voluntary insurance, or community financing (funding from households, the community, and the government).

By the 1970s risk-pooling mechanisms covered a remarkably large share of China's population relative to its still very low income level. But with the virtual disappearance of the rural cooperative medical system in the 1980s, China has become much more like the rest of the low-income world in insurance coverage: up to 700 million rural Chinese have lost their access to prepaid care and now pay out of pocket for almost all their health care. As incomes rise, so will demand to regain such coverage.

Health insurance coverage is still relatively high in urban areas, but the two urban insurance systems are in urgent need of reform. The government insurance system now covers about 30 million people, including current government workers, government retirees, the military, and university students. Its annual spending per member is three and a half times the average in China—389 yuan per member, compared with the national average of 110 yuan. The state enterprise insurance system covers an estimated 140 million employees and retirees of state enterprises. It spends about 259 yuan per member.

Rapid cost escalation has led to a fiscal crisis in both systems. The government's spending on its insurance system grew 15 percent a year in real terms from 1978 to 1993, and in recent years it has had to allocate additional funds to cover deficits. Some state enterprises have been unable to cover the health care bills of their employees and retirees. Since the systems are pay-as-you-go, state enterprises and government units with large numbers of retirees have particularly high costs.

Both systems also have large inefficiencies. Except for dependents in the state enterprise system, enrollees do not make significant copayments and therefore have few financial constraints on their consumption of medical services. In addition, care provided outside the systems is reimbursed on a fee-for-service basis, resulting in incentives for overprovision of services. And most important, health insurance coverage is tied to place of work, impeding the labor mobility essential for a modern economy.

In both rural and urban areas the government can play an important part in promoting the development of efficient and equitable risk-pooling mechanisms by adopting an appropriate policy framework. As new approaches to rural and urban insurance emerge, the government also needs to monitor and systematically evaluate them in order to adapt and improve this policy framework.

Options for rural reform. Because of public finance constraints, a rural health system financed largely from general revenue does not appear feasible in the medium term in China. (Government subsidies now cover only a small fraction of the costs of publicly provided services; about 85 percent of the costs are recouped from fees.) Nor are mandatory wage taxes feasible, because most of the rural population is self-employed. Community financing appears to be the most promising way to ensure universal, or near-universal, health coverage and efficient service delivery in rural areas without causing a major drain on government funds.

China has much experience with community approaches to rural health insurance, beginning with the commune-based rural cooperative medical system and including several ongoing community financing schemes that cover about 10 percent of the rural population today. A recent study of thirty poor Chinese counties, comparing villages that have community financing schemes with those that do not, showed that community financing is associated with higher use of health services at lower-level facilities, lower rates of morbidity, lower fees for primary care services, and a lower share of income from drug sales for township health centers and county hospitals (China Network and Harvard School of Public Health 1996; Jin 1995a). To the extent possible, coverage should be universal at the local level.

China's experience suggests that with appropriate government commitment, community financing is likely to be both administratively and financially feasible in many rural areas. Community financing has many advantages over the private, voluntary insurance that might develop in the richest rural areas. Private, voluntary insurance would exclude both the poor and those with high health risks and, if based primarily on fee-for-service payments, would also lead to cost escalation. Community financing also has many advantages over the present system, in which rural residents must pay out of pocket for services on a fee-for-service basis.

The government would need to monitor community financing approaches in different areas, evaluating their effect on health spending, efficiency, equity, and consumer satisfaction in order to inform medium- to long-term policy choices for rural health. These evaluations should focus on such key design features as the content of the basic benefit package, the size of the risk pool, management models, the level of copayments by type of service facility, ways to rationalize pharmaceutical use, and reimbursement methods for doctors, township health centers, and county hospitals.

Options for urban reform. Short- and medium-term measures are urgently needed to reform the government and state enterprise insurance systems.

- Management of insurance needs to be taken out of the state enterprises and government units in order to enlarge risk pools and achieve economies of scale in administration.
- Health insurance needs to be portable so that workers can move from one job to another without losing their benefits.

• The insurance programs need to move from a pay-as-you-go system to one in which contributions allow for expected expenditures in old age. (Or, alternatively, other mechanisms need to be developed to cover workers in old age.)
• Benefit packages need to be redesigned to be financially sustainable.
• Provider payment needs to move from fee-for-service to methods that help contain costs, such as capitated payment.

The State Council has sponsored experimental health insurance centers that are testing key reforms of the government and state enterprise systems. Experiments begun in December 1994 in the cities of Jiujiang and Zhenjiang provide for wage-based enterprise and employee contributions to individual and pooled accounts managed by the insurance centers. The State Council decided to expand these experiments to fifty more cities throughout China in 1996. All these experimental health insurance programs need to be carefully monitored and evaluated in order to inform policymaking. Government financing is needed for technical assistance, monitoring and evaluation, and strengthening of regulatory capacity. Many issues remain to be addressed, including how to pay for the unfunded health obligations of government and state enterprise workers (for example, for retirees of enterprises that will close) and what to do with state enterprise health facilities.

TABLE 1
Recommended health policy actions, 1997–2001

Objective	Short term (1–2 years)	Medium term (3–5 years)
Strengthen public health programs (chapter 3).	Provide full funding for the current Epidemic Prevention Service. Prohibit the agency from collecting user fees for public health services. Strengthen antitobacco programs, such as health education, bans on smoking in public places, and regulations on levels of tar and nicotine in tobacco products. Develop a strategic plan for public health, taking into account China's changing disease and risk patterns, to guide the work of the Epidemic Prevention Service and other agencies carrying out public health functions.	Increase funding for public health to about 6.5 billion yuan by 2001. Increase the tobacco tax by 20 percent or more to reduce smoking. Earmark the tax revenue to fund public health activities and health services for the poor. Upgrade the skills of staff in the Epidemic Prevention Service and other agencies involved in public health to carry out their new mandate. Ensure that priority public health programs reach the poor in urban areas, particularly the unregistered urban poor.
Ensure essential health services for the poor (chapter 4).	Phase in a program of geographically targeted subsidies for health services in poor villages in the 592 counties officially designated as poor. If a village has a viable community financing scheme, channel the subsidies through the scheme; otherwise, direct them to health providers. Phase in program-targeted subsidies for a limited number of health programs that particularly benefit the poor.	Continue to subsidize services for the poor and expand coverage. Monitor the effectiveness of subsidies to guide policy improvements. Over the long term, as more comprehensive urban insurance systems are set up, consider subsidizing the poor's contribution to such schemes.
Reform prices and provider payment mechanisms (chapters 5, 6, and 7).	Carry out a major study on price reform, focusing on the prices of health services and on ways to bundle services into packaged fees (this relates to reforms of the government and state enterprise insurance systems). Move away from reliance on fee-for-service provider payment methods. Test alternative provider payment methods that encourage efficiency in urban insurance experiments.	Gradually bring prices more in line with marginal costs, with periodic updates to account for inflation. Assist rural community financing plans in establishing provider payment mechanisms that encourage efficiency (such as salaries with performance bonuses for village doctors and capitation payments to county hospitals).

Over the medium to long term, urban health insurance coverage will also need to be broadened to include the increasing share of workers outside the state sector and their dependents, as well as the rapidly growing unregistered—or "floating"—urban population. And urban insurance centers will need to work toward harmonizing benefits, contributions, and portability across China.

Implications for public finance

The erosion in public health programs and the unmet demand for health services outside the insured population, especially among the rural poor, call for a substantial increase in health spending by the central government on the priorities discussed above. This report argues that spending on strengthened national public health activities should increase as a share of GDP by 2001. Public spending is also needed for new programs to provide essential services for the poor and to promote risk pooling in rural and urban areas.

China can afford these priority programs, and expected improvements in tax revenue will make them even more affordable. China's government budgetary expenditures are far smaller as a share of GDP than those of other countries. In 1994 these expenditures were 14.1 percent of GDP, and extrabudgetary expenditures were an additional 3.8 percent. Central govern-

TABLE 1 (CONTINUED)
Recommended health policy actions, 1997–2001

Objective	Short term (1–2 years)	Medium term (3–5 years)
Control investments and improve regional planning (chapter 5).	Improve and expand regional planning techniques developed under World Bank Health Loan III to better integrate various hospital systems. Change from centrally managed staffing of individual health facilities to nationwide and provincewide human resource planning to support the broad health reforms proposed here.	Institute nationwide regional planning guidelines. Develop government oversight mechanisms to avoid oversupplies of physicians (particularly specialists), tertiary hospital beds, and expensive diagnostic equipment.
Promote efficient risk pooling in rural and urban areas (chapters 6 and 7).	*Rural areas* Develop national guidelines for community financing that address the size of the risk-pooling base, management structure, and community control mechanisms. Organize provincial technical assistance teams to assist community programs. Implement pilot programs.	Provide training, technical assistance, and modest central government subsidies to rural communities that establish community health financing. Monitor the impact of alternative forms of community financing on health status, access, organization and delivery of services, pharmaceutical use, health spending, and so on.
	Urban areas Extend urban insurance experiments to other provinces and systematically evaluate these additional experiments. Consider financing some of the start-up costs of new schemes (such as information systems, capital equipment, and training) with government assistance.	Gradually extend social health insurance to cover workers employed by joint ventures, smaller collective industries, and private enterprises. Mandate open access on a voluntary basis to all urban residents not covered by their employers. The government could eventually finance insurance for the indigent. Experiment with risk pooling in larger areas, such as by province. Reform provider payment methods. Insurance centers could move toward negotiating capitation contracts with providers covering all levels of services, under risk-adjusted capitation rates. Alternatively, they could move toward a payment system for hospital services based on diagnosis-related groups. Prices should be based on reasonable costs, with periodic adjustments as needed.

ment expenditures as a share of all government expenditures are also unusually low—only 40 percent, compared with an average of 78 percent in other developing countries.

The World Bank projects that China's GDP will double between 1993 and 2001, from about 3,450 billion yuan to 7,500 billion yuan (in 1993 prices). It also projects that government revenues will rise significantly as a share of GDP during the Ninth Five-Year Plan (1996–2000) if China implements suggested changes in tax administration and structure. With these changes, tax revenues should increase by an amount equal to 6 percent of GDP. The priority health programs recommended in this report might cost about 13 billion yuan by 2001, less than 1 percent of projected government revenue. Most of this spending—90 percent—would be for public health activities and subsidized services for the poor.

A related recommendation is to increase the tax on tobacco by 20 percent or more. A 20 percent increase in the tobacco tax is projected to generate 10 billion yuan in additional revenue annually, revenue that could be used to help finance the public spending increases recommended in this report. By reducing smoking, this tax would also produce important health benefits.

Experience in high-income countries suggests that China is at a critical juncture for redirecting its health policies. If it adopts the package of policy reforms recommended in this report, it could expect, within ten to fifteen years, to achieve much higher levels of prepaid health coverage, to eliminate most of the excess disease burden among the poor, to have maintained the general improvements in life expectancy, and to have stabilized health expenditures at 5–7 percent of GDP (just below the range for OECD countries). Failure to adopt these policies would risk leaving a large share of the population without health insurance, jeopardize health improvements, and encourage growth in health spending to 10 percent of GDP or more (as in Argentina, France, and the United States).

The choices that China makes in health financing policy in the coming years will rest not only on financial and economic analyses. These choices hinge fundamentally on the judgments China makes about what kind of society it wishes to be and what value it places on social cohesion, poverty alleviation, equity, consumer choice, and quality of care. This report argues that, in the right policy environment, achieving these broader social goals can be consistent with measures that improve economic efficiency in the health sector.

chapter one

Assessing the Performance of China's Health Care System

A nation's health policies directly affect both the health of its population and the operation of its health care system. This chapter sets up the analytical framework for evaluating health finance policies in China by assessing the performance of China's health sector and examining the health policy issues it raises.

The assessment of China's health sector uses four broad measures of performance. The first is *health status*. The health status of China's population has improved enormously since 1949. A good indicator of this is life expectancy: In 1990 a typical country at China's income level achieved a life expectancy of about sixty-four years, while China's was sixty-nine. As discussed below, however, the trend in under-five mortality appears to be a cause for concern.

The other three performance measures—*financial access to health services, efficiency,* and *total cost*—relate to the financing and provision of health services. Policies ensuring broad financial access to health services help ensure that health care is delivered equitably. This desirable goal is best achieved through risk-sharing (insurance) mechanisms that provide prepaid coverage for a reasonable range of services. The percentage of a population covered by such risk-sharing mechanisms is therefore an important indicator of a health system's performance. Health policies also affect the efficiency of care—whether services are produced at the least possible cost and whether spending is efficiently allocated (producing value for money).

The total cost of the health resources used by a nation is another important policy outcome. As experience around the world suggests, spending more on health is not always required to improve health outcomes: some high-spending countries (such as the United States) get low returns relative to their resource commitments. Another dimension of cost is the indirect economic losses that result from a health care system. For example, insurance systems that tie health insurance coverage to certain employers may hamper labor mobility.

Health policy choices are not the only influences on a nation's health status and the operation of its health system. Factors outside the health sector are also important. Income and education levels, for example, form the foundation for a nation's health policies. As income increases, so does the ability to acquire the necessities for good health—adequate food, clean water and sanitation, satisfactory shelter, and access to health services. Similarly, as education levels rise, so does the ability of the population to make informed choices about health, income disposal, and personal behavior.

Demographic changes are another important influence on the health sector. China's demographic patterns have changed dramatically over the past four decades. Rapid declines in fertility and mortality, the aging of the population, and the potential for explosive urban growth affect both health conditions and planning for the evolution of the health system and its finance (Jamison 1996).

Health status—progress and problems

In 1984 the World Bank's first health sector report on China called for completing the first Chinese health care revolution: extending successful programs for improving child health and controlling endemic infections into poor rural areas; consolidating and deepening the health gains achieved in most of rural China by reversing the breakdown in cooperative medical services; and seeking new ways to finance public health programs that were being neglected by local providers embracing fee-for-service practices (World Bank 1984). The report also encouraged a second revolution: preventing and managing the growing burden of noncommunicable diseases at a much lower cost than in the high income countries, where such diseases had emerged earlier as a dominant problem. The World Bank's second health sector report on China dealt much more extensively with noncommunicable diseases and their risk factors (World Bank 1992a). This section reviews the status of two of these issues—child health and the noncommunicable disease burden—in China today.

Slowing improvement in child health

Despite rapid income growth in the past decade, China's progress in improving child health appears to be a cause for concern. Analysis in an earlier World Bank report suggested that the infant mortality rate stopped declining in 1982 (World Bank 1992a, pp. 6–7). A later overview points to recent unexpected outbreaks of immunizable diseases in some areas (Parker n.d.). This overview also presents evidence from surveys in nine provinces that a key indicator of child malnutrition—the percentage of children with very low height for their age—increased in rural areas between 1987 and 1992, although malnutrition in urban areas of those provinces declined sharply.

To assess under-five mortality, this report commissioned a complete analysis of national trends using recently available census and survey data (Hill and Maeda 1997). While this analysis could also have looked at infant mortality, demographers have concluded that estimates for under-five mortality are consistently more robust and reliable. UNICEF regards the under-five mortality rate as the best indicator of social development because it accounts for the health and knowledge of the mother, immunization levels, use of appropriate health services, access to water supplies, sanitation conditions, and the overall safety of the child's environment (UNICEF 1989, p. 82).

The analysis concludes that the under-five mortality rate in China declined steadily until the early 1980s and then stagnated until 1991. But Chinese researchers question the reliability of these child mortality estimates from censuses and fertility surveys and consider death registration data in China to be a more reliable source. (International demographers do not commonly use death registration data to estimate child mortality in developing countries, preferring to derive mortality estimates by applying indirect estimation techniques to censuses and surveys, as was done for the estimates in table 1.1.) Based on death registration data, under-five mortality in China declined in the 1990s—from 61 per 1,000 live births in 1991 to 51 in 1995.

Experience in other countries suggests that the under-five mortality rate need not plateau as China's did in the late 1980s (as measured using census and survey data). Sri Lanka's per capita income is slightly higher than China's and its 1975 under-five mortality rate was moderately lower, but by 1990 its under-five mortality rate had dropped to half the rate in China (table 1.1). Japan's infant mortality rate in 1951 was about the same as China's in 1976, but it then dropped by a third in six years. Indeed, until 1951 the decline in Japan's infant mortality rate was remarkably similar to that in China twenty-five years later, but Japan's decline continued and no plateau was observed (Parker n.d.).

Projected growth in the noncommunicable disease burden

The old and the young are afflicted by very different health problems. Noncommunicable diseases—stroke, cancer, ischemic heart disease, and chronic lung disease—account for most mortality in people in late middle age and older age groups. A relatively small number of infectious diseases, most of which are inexpensive to prevent and treat, cause most deaths of children. In China, as in other countries, noncommunicable diseases are projected to account for an increasing share of the disease burden (figure 1.1). The contribution of injuries is expected to change little, while infectious diseases are expected to steadily decline in importance (see Murray and Lopez 1996).

Much of the projected increase in the importance of noncommunicable diseases results from unalterable demographic changes. But part comes from controllable risks. Tobacco use is the most important example. In 1990 tobacco use accounted for about 800,000 of the 8.9 million deaths in China, and projections of the effects of past and future tobacco use suggest that more than 2 million tobacco-related deaths will occur in 2020. Unless tobacco use can be curtailed, tobacco-related deaths will have almost tripled between 1990 and 2020 and almost doubled as a share of all deaths (Murray and Lopez 1996).

Enormous resources can be devoted to preventing and treating noncommunicable diseases through interventions that are costly and of limited efficacy. But the right incentive environment can encourage experimentation with and adoption of more cost-effective approaches. This report points to ways for improving China's decisions about preventing and managing noncommunicable diseases (see chapter 3).

Nonhealth indicators of the health care system's performance

This section highlights trends and policy issues associated with three nonhealth outcomes of health policy—access, efficiency, and cost of health care.

TABLE 1.1
Under-five mortality rate in China and other Asian economies in selected years, 1960–90
(deaths per 1,000 live births)

Year	China	Hong Kong	India	Indonesia	Japan	Sri Lanka	Vietnam
1960	173	53	235	214	37	140	105
1965	144	n.r.	n.r.	n.r.	n.r.	n.r.	n.r.
1970	115	n.r.	n.r.	n.r.	n.r.	n.r.	n.r.
1975	85	17	195	151	11	69	68
1980	60	n.r.	n.r.	n.r.	n.r.	n.r.	n.r.
1985	44	n.r.	n.r.	n.r.	n.r.	n.r.	n.r.
1990	44.5	7	127	111	6	22	46

n.r. Not reported.
Source: For China, Hill and Maeda 1997; for other economies, World Bank 1993b.

FIGURE 1.1

The growing burden of noncommunicable diseases

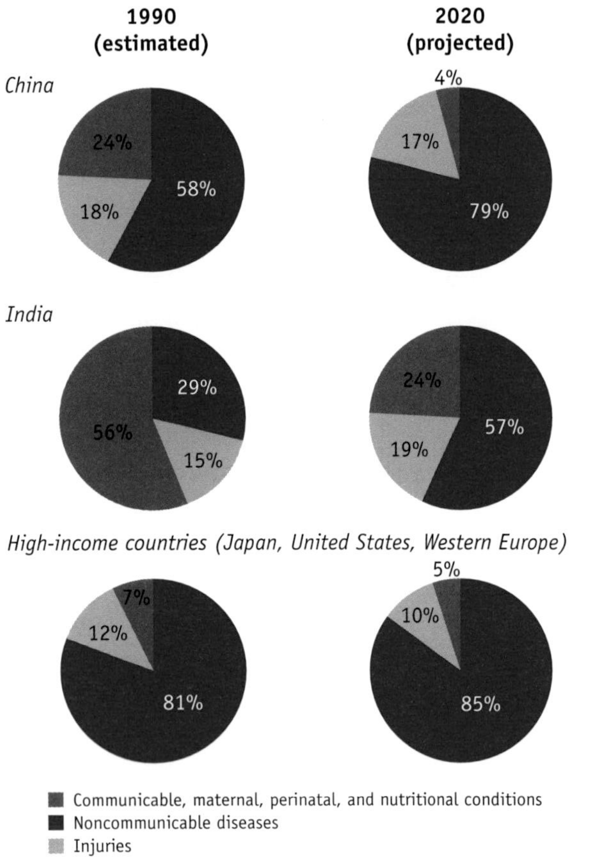

Source: Murray and Lopez 1996.

New issues in access and equity

People in China have relatively good physical access to basic health care services. High population density and a well-developed health infrastructure mean that geographical barriers are modest for all but a significant minority living in mountainous or remote rural areas. Until recently the government made cost-effective public health services widely available, minimizing financial barriers to these services. And the cost of routine, basic outpatient health services is low enough so that most nonpoor households can pay for them out of current income or savings.

Catastrophic care poses more of a problem, however, because it involves services that are expensive relative to household income. Many households must either forgo treatment or go deeply into debt to pay for it. As much as 70 percent of total health spending in many countries goes to catastrophic care. An important performance indicator for a nation's health system, then, is how efficiently it finances and provides these services. What can be included under catastrophic care coverage will vary with a country's income level and health infrastructure.

Since only a fraction of the population needs catastrophic health care in any year, pooling risks is the best mechanism for financing these services—under a system financed by general tax revenue, social insurance, or private voluntary insurance. The need for risk pooling is an issue both of equity, since the poor will require subsidies, and of efficiency, since all but the very wealthy (who can be self-insured) will generally benefit. Low-income countries typically lack the institutional and financial capacity to offer risk pooling to most citizens. But as incomes rise, risk pooling typically benefits a growing share of the population—often, at least initially, through straightforward government or collective finance of clinics and hospitals open to everyone.

From the late 1960s through the early 1980s China provided an exceptionally large share of its population (relative to its income level) with at least some risk pooling, although not without many problems (see chapter 6). In 1981 only 29 percent of China's population had no risk-pooling coverage (figure 1.2). But by 1993 the uninsured had grown to almost 80 percent of the population, largely as a result of the fundamental changes in China's economy. The challenge is to restore broad access to health care in the new economic environment.

A mounting problem of inefficiency

That total health spending has been growing rapidly in China—even as some key indicators of health status have not improved—suggests that China's health sector faces a mounting problem of inefficiency. The allocation of public spending favors less cost-effective hospital services over highly cost-effective public health activities. Distorted prices encourage the overuse of drugs and high-technology tests. Fee-for-service payment encourages overprovision. And multiple vertical health delivery systems have led to excess capacity and waste. Reversing these trends will require reallocating resources, both public and private. Risk-pooling and provider compensation arrangements need to be designed to contain costs, extend access, and promote greater quality of care and value for money.

FIGURE 1.2

A big jump in China's uninsured population

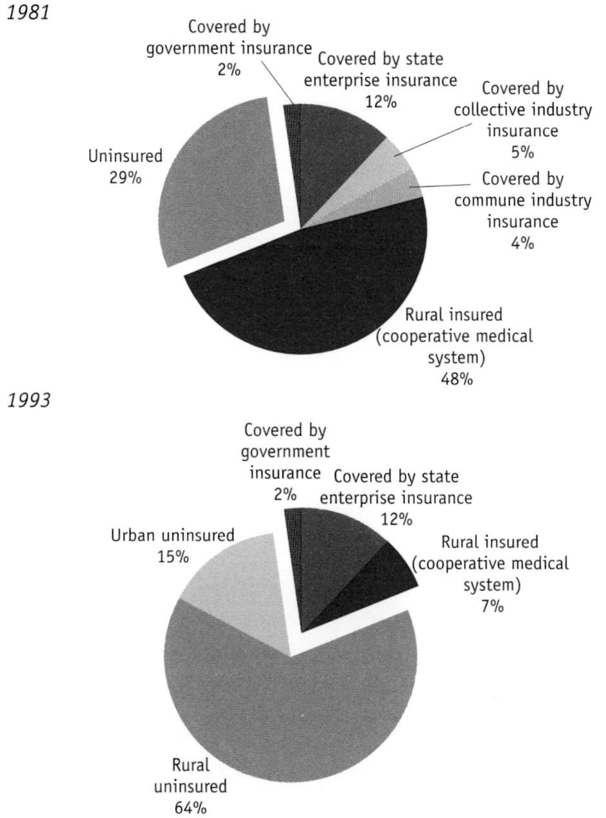

Source: World Bank 1984; Wei 1996.

FIGURE 1.3

Varied performance in containing health care costs

Health spending as a percentage of GDP, 1960–93

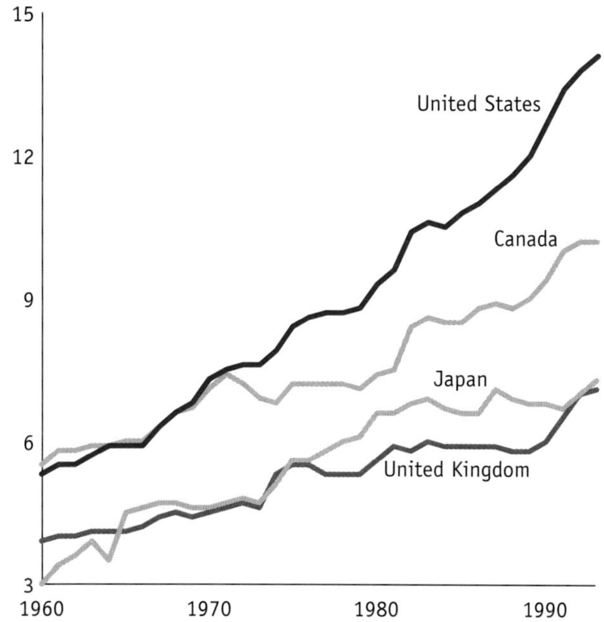

Source: OECD 1995.

Rising cost of health services

Demographic change and economic growth virtually guarantee that health expenditures in China will grow as a share of GDP. But policy choices can determine whether the growth in spending is excessive and whether it efficiently expands access and improves health outcomes.

Some countries have done far better than others in controlling health care costs (figure 1.3). In 1960 Canada, Japan, the United Kingdom, and the United States all spent 3–5.5 percent of national income on health, a share similar to what China spends today. But spending rates diverged sharply over the next thirty years, with health spending reaching 14 percent of GDP in the United States by 1993—even though 15 percent of its population still is uninsured. By contrast, Japan spent only 7.3 percent of GDP on health in 1993, with nearly universal coverage and the world's highest life expectancy.

Perhaps the most important factor outside the health sector that leads to rising costs is the aging of the population and the accompanying epidemiological changes. The World Bank has concluded that the aging factor alone will increase health spending in China from about 3.2 percent of GDP in 1992 to 5 percent in 2010 and 7 percent by 2030 (World Bank 1992a).

Factors in the health sector affect costs even more, however. Between 1978 and 1993—when China's health expenditures grew an average 10.9 percent a year—factors exerting upward pressure on expenditures included the tendency for the use of health services to rise faster than income (a pattern repeated worldwide) and the shift from salaried to fee-for-service compensation of providers. Distorted prices and the profitability of drugs and high-technology diagnostic tests also put upward pressure on health care costs (see chapter 5).

During the same period the coverage of risk-pooling arrangements in rural areas declined sharply, from 48 percent in 1981 to 7 percent in 1993 (see figure 1.2). Under declining coverage, the willingness and ability of patients to pay out of pocket for services limit providers'

TABLE 1.2
Recommended health finance policies and their potential impact on health sector performance

Recommendation	Improving health status	Promoting access and equity	Improving efficiency	Containing costs
Fully fund key public health programs	↑↑	↑	↑	↑
Subsidize essential health services for the poor	↑↑	↑↑	↑	•
Reform prices and provider payment mechanisms	•	•	↑↑	↑↑
Control investments in capital and human resources and improve regional planning	•	•	↑↑	↑↑
Promote universal risk pooling	•	↑↑	•	•

• denotes no significant impact; ↑ denotes moderate impact; ↑↑ denotes strong impact.

capacity to overprescribe or even to supply needed services. No such constraint operates under urban insurance programs. Under the government health insurance system, for example, the government reimburses providers for essentially all the patient procedures that the system approves. The differences between the rural and urban incentive regimes had clear consequences: between 1981 and 1993 per capita health spending in rural areas increased from 21 yuan to 60, while the government insurance system's per capita spending increased from 96 yuan to 389.

There are many ways to achieve the efficiency gains of broad risk pooling without creating incentives that lead to excessive cost escalation. But the government insurance system embodies incentive arrangements like those that have led to excessive growth in costs in such countries as the Republic of Korea, Singapore, and the United States. And as incomes grow in rural China, so too will the demand for risk pooling and prepaid care. Relying on out-of-pocket financing to keep costs down is not only undesirable because of the efficiency losses, but also impractical in the face of probable demand for prepaid arrangements.

In the medium term, health financing policies also need to minimize indirect costs to the economy (principally distortions in the labor market) by, for example, separating the provision of health services from employment.

Conclusion

China faces some disturbing trends and challenging policy issues in several areas of health sector performance reviewed in this chapter. Data suggest that the child mortality rate may not have improved, although this finding is contradicted by death registration data and must be seen as tentative. Risk pooling, or insurance coverage (including the rural cooperative medical system), and access to care have markedly declined. At the same time, real expenditure per capita has increased by a factor of more than 2.5 in rural areas and as much as 4 in urban areas since 1978. Financial incentives for providing preventive care have eroded, while financial incentives (and opportunities) for providing excessive or inappropriate care have multiplied. And initiatives for the collective action required for efficient risk pooling have not been widely or consistently implemented.

The economic reforms begun in China in the late 1970s have brought rapid economic growth, but they have also had unintended and sometimes detrimental effects in the health sector. This report lays out options for adapting China's health policy to the new economic environment. The following chapters recommend policy measures to deal with each of the challenges reviewed in this chapter (table 1.2).

chapter two

Health Services and Their Financing

To provide health services to its population of 1.2 billion, China has some 200,000 health establishments and a wide array of supporting research organizations. The country has 5.3 million health professionals, who make up about 0.8 percent of the labor force. They include 1.9 million doctors (about 1.6 doctors per 1,000 people) and 1 million nurses. Doctors are trained to join one of three categories: junior doctors (19 percent of the total), senior doctors (62 percent), and doctors of traditional Chinese medicine (19 percent). Senior doctors are concentrated in medium-size and large cities. Village doctors (also known as barefoot doctors), who have much less training, are excluded from these estimates and categories.

China has some 3 million hospital beds, 2.4 per 1,000 people. This ratio is higher than that in Africa and the rest of Asia and nearly as high as that in Latin America and the Caribbean. Only the OECD and Eastern European countries have significantly more beds per capita. In 1994 China's bed occupancy rate was 69 percent, and the average hospital stay was fifteen days (China, State Statistical Bureau 1994; World Bank 1993b; China, Ministry of Health 1993b).

Most hospitals are part of the Ministry of Health system, including its provincial and county affiliates, or are operated by state enterprises. Others are run by village and township collectives. And private practitioners operate an estimated 161,000 clinics in urban areas.

More than half of China's health workers are employees of the Ministry of Health or its provincial health bureaus. At the end of 1993 the Ministry of Health and its provincial affiliates employed 1.7 million hospital workers, 1.1 million health workers in township health centers, 250,000 workers in the Epidemic Prevention Service, and a little less than 100,000 in maternal and child health programs. The Ministry of Health also finances the education of 220,000 medical students enrolled at 120 medical schools. Thirty of these schools belong to the ministry; others belong to local or provincial governments.

China's state enterprises employ another 1.4 million health workers and operate 700,000 hospital beds, both accounting for roughly a quarter of the national total. Since the 1950s state enterprises have provided health services directly to their employees (including retirees) and their employees' families. Most of their health facilities are small clinics and health posts, but some are hospitals and large service centers. Some of the large hospitals and clinics serve third parties on a fee-for-service basis. The Taiyuan Machinery Works in Shanxi Province, for example, owns and operates a 300-bed hospital that receives a quarter of its revenue from patients not affiliated with the enterprise. In 1993 state enterprise health facilities delivered 18 percent of outpatient and emergency care in China and 13 percent of inpatient treatment. The proposed reforms of state enterprises would separate such services from their regular business. But some enterprises may prefer to expand their medical business rather than give it up. The Taiyuan Machinery Works, for example, plans to expand its hospital to 500 beds.

Outside the Ministry of Health and state enterprise systems, thousands of health workers are employed by other government institutions, such as the military and prison systems. In addition, there were some 150,000 health workers in private practice in 1990, and an estimated 190,000 in 1993 (excluding village doctors).

The three tiers of the rural health delivery system

The rural three-quarters of the Chinese population is served by a three-tier system of health services and referral. Farmers and their families normally enter the system through a visit to one of China's approximately 1.33 million village health workers (955,000 village doctors and 375,000 health assistants). These health workers, who work independently, engage in both health care and farming and often earn as much from farming as from medicine. Many received rudimentary training as barefoot doctors in the 1960s and 1970s and continue working in the villages of their birth.

In this first tier of the system village doctors diagnose and treat patients, prescribe pharmaceuticals, and refer patients to higher levels of service as warranted. Village doctors generally operate on a fee-for-service basis, but they also depend for income on the markup on drug prescriptions (typically about 15 percent).

The village doctor may refer patients needing a higher level of care to the nearby township health center or hospital, the second tier of the rural health care system. There are some 52,000 rural township health centers, operating 730,000 beds, about a quarter of all hospital beds in China. Township health centers are staffed by junior doctors and other medical personnel and can deliver babies, treat infections and wounds, and provide basic surgery such as appendectomies. The health centers depend mostly on patient fees but also receive subsidies from local governments that cover part of their costs. These facilities tend to have lower bed occupancy rates than the higher-level hospitals. Some farmers referred for hospital attention bypass the township facilities because they question their quality, and go directly to the county hospitals.

There are about 4,000 county hospitals in China, making up the third tier of the rural health care system. These hospitals are usually the last point of referral for inpatient treatment of rural residents, since few farmers can afford treatment at specialized, big-city hospitals. County hospitals have on average about 300 beds. They typically have five departments—obstetrics and gynecology, pediatrics, general surgery, internal medicine, and laboratories and x-rays—as well as emergency room facilities.

Independent of this three-tier system in rural China are three important vertical public health services: the

Epidemic Prevention Service and the Maternal and Child Health Program, both under the Ministry of Health, and the Family Planning and Reproductive Health Program, under the Family Planning Commission. These programs receive budgets from provincial and county governments and also collect fees from their clientele. User fees supplement the public funding to a greater or lesser degree depending on local economic conditions. In wealthier provinces, such as Jiangsu, the programs are largely self-supporting and receive any supplementary financing needed through rural collective enterprises. In poorer provinces, such as Shanxi and Guizhou, where clients and enterprises are less able to reimburse the programs, they depend more on government finance.

FIGURE 2.1

Growing health spending from most sources in China

Billions of 1993 yuan

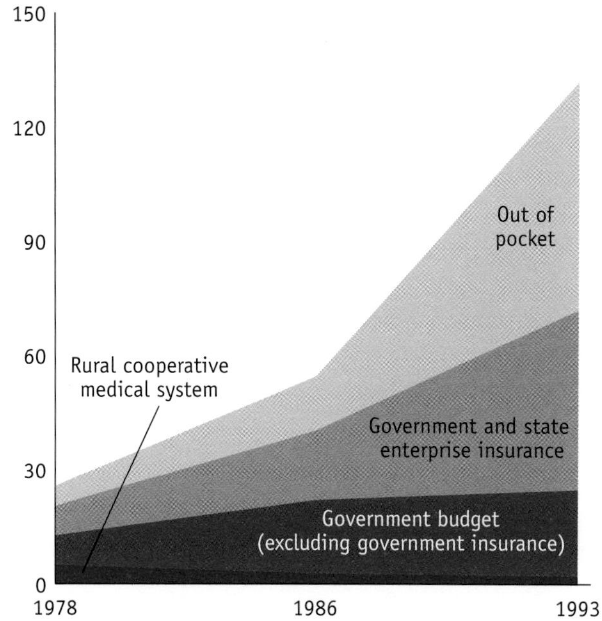

Source: See annex table A.3.

Sources of health spending

China allocated about 3.8 percent of GDP to health in 1993, the most recent year for which comprehensive national health statistics are available.[1] It spent somewhat less on health care in earlier years—an estimated 2.9 percent of GDP in 1978 and 3.0 percent in 1986. The most dramatic change in health financing between 1978 and 1993 was the decline of the rural cooperative medical system (figure 2.1). The financing gap that resulted was filled mainly by private out-of-pocket spending. This shift from cooperative to personal financing may have hit the poor hardest, since they are among the least able to pay for health care out of savings.

Several other significant changes occurred:

- The share of health spending from the government budget (excluding subsidized care for government workers provided through the government insurance system) declined from 32 percent to 14 percent between 1986 and 1993.
- The share of spending contributed by the cooperative medical system fell from 20 percent in 1978 to 2 percent by 1993.
- Out-of-pocket payments rose from 20 percent of the health sector's revenue in 1978 to 26 percent in 1986 to 42 percent in 1993, transforming its financing base.
- Government health spending (excluding the government insurance system) almost tripled in real terms between 1978 and 1993. But private health spending grew even more rapidly, increasing by a factor of ten. Spending by the government and state enterprise health insurance systems rose only slightly as a share of national health expenditures, from 30 percent in 1978 to 33 percent in 1986 to 36 percent in 1993.

Uses of health spending

The bulk of China's health sector funds—public, private, and insurance-based—is absorbed by hospital services and the purchase of drugs. In 1993 three-quarters of China's health spending went to pay for inpatient or outpatient hospital care, with about 60 percent of that spending going to pharmaceuticals (table 2.1). Public sector staff and institutions deliver most health services, collecting fees from patients or their employers. Hospitals are expected to cover 85 percent or more of their costs from patient revenues. Most manage to balance revenues and expenditures.

Chinese public hospitals are fairly autonomous compared with those in many other countries. The government sets basic salaries, but the hospitals determine bonuses, which can be one to two times the basic salary. Hospitals can move funds across budget categories and make their own equipment purchasing decisions (though capital investment plans need the approval of

TABLE 2.1

Sources and uses of health financing, 1993

(millions of 1993 yuan)

	Source of finance					
	Government budget[a]			Insurance		
Use of finance	Health recurrent expenditure[c]	Traditional Chinese medicine	Other	Government	State enterprise	Community financing[d]
Hospitals						
Outpatient	1,912	349	0	1,266	3,737	737
Inpatient	2,571	445	0	2,569	7,583	139
Total	4,483	794	0	3,835	11,320	876
Pharmaceuticals						
Hospitals	0	0	0	7,639	22,552	1,367
Individual providers, retailers	0	0	0	0	0	0
Testing	112	0	0	0	0	0
Epidemic Prevention Service	1,305	0	0	0	0	0
Maternal and Child Health Program	324	0	0	193	569	0
Other primary health care	1,344	78	0	0	0	0
Family planning	0	0	2,292	0	0	0
Medical research and education	408	42	1,160	0	0	0
Construction	2,810	3	1,144	0	0	0
Other	0	0	2,579	0	0	0
Total	10,786	917	7,175	11,667	34,441	2,243
Percentage share	8.2	0.7	5.4	8.8	26.1	1.7

a. Excludes government insurance system.
b. In the Chinese health accounts data, society financing can be either public or private. For example, for the Epidemic Prevention Service and Maternal and Child Health Program, society financing from others refers to user fees.

the appropriate planning commission). But they have little autonomy in personnel decisions. Personnel are assigned to hospitals by the health bureaus according to the staff quota set by personnel bureaus at the same administrative level.

China devotes more of its health spending to pharmaceutical purchases than most low-income countries—52 percent in 1993. In OECD countries spending on drugs averages 14 percent. About 85 percent of pharmaceutical sales in China in 1993 occurred in hospital inpatient or outpatient settings.

Overprescribing and the misallocation of resources toward drugs are a major efficiency issue in China's health system. Other problems are overuse of high-technology diagnostics and the long average hospital stay (fifteen days). These efficiency issues are discussed in more detail in chapter 5.

Government recurrent spending

Spending under the government's recurrent health budget—by the Ministry of Health and related departments at the provincial and county levels—amounted to about 8 percent of all health spending in 1993. Recurrent spending includes that on public hospitals (40 percent of the total), the Epidemic Prevention Service (12 percent), and maternal and child health services (3 percent). Public spending on family planning, which is outside the Ministry of Health's recurrent health budget, amounts to about 1.7 percent of health spending. Public spending on traditional Chinese medicine facilities (also separate from the recurrent health budget) is smaller yet, at 0.7 percent.

Public spending on the health care of government employees and related groups (8.8 percent of total health spending), though part of government health spending, is not included in the recurrent health budget. (OECD countries generally keep this account separate, as part of government nonwage employee compensation.) Payments under the government insurance system rose from 14 percent of government health spending in 1978 to 46 percent in 1993. Thus the bulk of the 200 percent increase in government health spending went to pay the health care costs of government employees.

Source of finance								
	Society financing[b]				Out of pocket			
Government and NGOs	Enterprises	Villages	Other private	Other	Urban	Rural	Total	Percentage share
1,276	0	0	0	0	5,828	9,459	24,564	
0	0	0	0	0	1,096	1,170	15,573	
1,276	0	0	0	0	6,924	10,629	40,137	30.4
0	0	0	0	0	10,812	17,844	60,214	45.6
0	0	0	0	0	3,300	5,417	8,717	6.6
0	0	0	0	103	0	0	215	0.2
0	137	195	0	1,604	0	0	3,241	2.5
0	137	195	0	534	381	799	3,132	2.4
0	0	0	0	610	0	0	2,032	1.5
0	0	677	0	0	0	0	2,969	2.2
0	0	0	0	816	0	0	2,426	1.8
342	877	0	143	1,067	0	0	6,386	4.8
0	0	0	0	0	0	0	2,579	2.0
1,618	1,151	1,067	143	4,734	21,417	34,689	132,048	100.0
1.2	0.9	0.8	0.1	3.6	16.2	26.3		

c. Health recurrent budget includes the Ministry of Health and local government public health departments. Most spending is at the local, not central, level.
d. Cooperative medical system.
Source: Wei 1996.

There was also a big increase for the high-priority family planning program. These changes have shifted funds away from preventive health services.

China's Epidemic Prevention Service programs, which were entirely subsidized before 1985, are now among those being pushed hardest to finance their work through user fees. For example, while government regulations require restaurant workers to be tested by the Epidemic Prevention Service, their employers must pay for the tests. Shanxi's Epidemic Prevention Service was entirely financed by the provincial government until 1991. But by late 1994 only 65 percent of its income came from the provincial treasury, and the rest from fees for testing services in cities and towns. In villages, where costs cannot be as easily recovered, the frequency of field visits has declined significantly. The Epidemic Prevention Service of Guizhou, a relatively poor province, has cut its village fieldwork even more than wealthier provinces. In Guizhou most of the program's income still comes from the government because there is little capacity to sell testing services even in cities and towns (Jin 1995a, pp. 25–26).

Capital spending

Capital spending patterns in the health sector in China, as in many countries, show scope for improving the allocation of resources. A serious and long-standing issue is duplication of facilities. In urban areas there is overlap among the Ministry of Health, state enterprise, and traditional Chinese medicine facilities. In rural areas there are duplication and overlap of services among maternal and child health centers, family planning services, township health centers, and epidemic prevention stations (see box 5.3).

From 1985 to 1989 as much as 80 percent of health investments went into hospital construction and equipment, while less than 10 percent supported public health and high-priority basic clinical services. In the Eighth Five-Year Plan period (1991–95) the central government established a special fund of 1.1 billion yuan to strengthen public health and basic health care, known as the Three Items Construction Program. To participate, local governments had to provide complementary funds for epidemic prevention and maternal and child health care programs at the provincial and county levels (Hou and Zhou 1995). The idea was sound, but the poorest counties and townships, unable to generate counterpart funds, have been the least able to benefit from the program.

Resource allocation is also an issue in public hospitals' capital investment decisions. Hospitals' reputations and their ability to attract clients now depend on having high-technology equipment—computerized tomography (CT) scanners, ultrasound, and other diagnostics have come to symbolize satisfactory health care to Chinese consumers. At the end of 1993 China had 1,300 CT scanners, 200 magnetic resonance imaging (MRI) machines, and 1,200 color Dopplers (Hu Haobo 1995). Many specialists believe that these are not the most cost-effective investments for a country at China's stage of development—or even for much wealthier countries.

Investments in health research can do much to help address China's remaining health sector problems, and government support of research is an essential element of public health policy (box 2.1). The Chinese national health accounts data do not disaggregate investments in medical research, but include them in the estimate for medical education and research, which totals about 1.8 percent of all health spending.

Fiscal barriers to bigger and more equitable health budgets

Two fiscal problems complicate the government's efforts to finance health services and to promote redistribution between rich and poor areas of the country. First, revenues are decentralized, limiting the central government's ability to transfer resources from rich to poor provinces. Only about 4 percent of the total recurrent health budget in 1993 fell under the direct control of the central government (Berman and others 1995, p. 28). Province, prefecture, county, and township spending accounted for the rest. This decentralization hampers special assistance for the poor, since they generally live in provinces with limited capacity to tax and to redistribute benefits through publicly subsidized health services. Furthermore, regional income disparities are growing.

Second, government revenue has been steadily declining as a share of GDP. In concert, China's overall budgetary expenditures declined from 33.8 percent of GDP in 1978 to 13.8 percent in 1994 (World Bank 1996b).

BOX 2.1

Health research and development: A neglected part of China's health system?

The high-income countries of the OECD invest 3–4 percent of their health expenditures in research and development (R&D)—developing countries, China among them, 0.5–1 percent. OECD spending on R&D splits about evenly between research (mostly public sector) and product development (mostly private sector). In China, as elsewhere, elite universities and specialized institutions (such as the Chinese Academy of Preventive Medicine) undertake most health research. These institutions, along with industry, also engage in product development, including pharmaceuticals. But there are almost no newly registered drugs from China or, indeed, from any developing country.

Should China consider investing more of its health budget in R&D? And if so, where? One argument is that since the OECD countries invest so heavily in R&D, the rest of the world need not. A recent report of the World Health Organization (1996) advances a different argument, suggesting four broad areas where China might productively invest R&D resources:

- Operational research focused on control of the diseases of extreme poverty that still affect perhaps 100 million Chinese.
- Biomedical research and new product development to counter infections that are still evolving (such as AIDS or drug-resistant tuberculosis) with better vaccines, drugs, and diagnostic tools.
- Epidemiological, preventive, and clinical research to address the rapidly increasing problems of noncommunicable diseases and injuries with interventions inexpensive enough to be widely implemented and sustained.
- Health policy and systems research to address issues of cost containment, access, and quality of service.

The case for expanding China's efforts in these four areas appears strong. The WHO report also stresses the importance of competitive allocation of R&D resources and full engagement with the international R&D communities to help China avoid repeating work already completed elsewhere.

Given the importance of new knowledge for improving health outcomes and constraining cost growth, China may wish to undertake an in-depth review of its health R&D. This review could be carried out by a committee like the one that prepared the WHO report—one of whose members was a Chinese scientist. While most members of such a committee would of course be from China, it might also be useful to include a few eminent outside scholars and industrial scientists for the different perspectives they would bring.

These problems of decentralized spending and limited resource mobilization are addressed in a recent World Bank report (1996b). The report projects that increasing tax enforcement, broadening the tax base, and taking some tax policy steps could raise government revenue by an amount equal to 6 percent of GDP by 2000 (see table 8.4). That would mean an increase in the central government's share of revenues from about 40 percent in 1994 to 60 percent, a shift that would ease fiscal transfers to poor provinces and counties. A key challenge in intergovernmental fiscal relations is to design and implement a grants scheme to redistribute the central government's revenue surplus to the poorer provinces.

Who gets health services—and how do they pay for them?

Two major concerns in China's health sector are the declining health insurance coverage (under the government and state enterprise insurance systems and the rural cooperative medical system) and the rural poor's inadequate access to health services. Health insurance coverage and health spending differ markedly among population groups (figure 2.2). In 1993 risk-pooling mechanisms (the government and state enterprise insurance systems and rural community financing schemes) covered only 21 percent of the population, but accounted for 38 percent of health expenditures.

Among the poorest fourth of the rural population (the poorest fifth of all Chinese), virtually none has prepayment or insurance arrangements to ensure funding for health services. This group accounted for only about 5 percent of health spending in 1993—clear support for the view that the rural poor receive an inequitably small share of available health services (see chapter 4).

Urban dwellers, who account for about 53 percent of earned income, were the beneficiaries of at least two-thirds of all health spending. The high income elasticity of demand for health services explains only part of this disparity.

Health coverage in urban areas

Official data indicate that 30 million people, or some 2 percent of the Chinese population, are eligible beneficiaries of the government health insurance system.

FIGURE 2.2

Widely varying health spending among groups in China

Per capita health spending, 1993

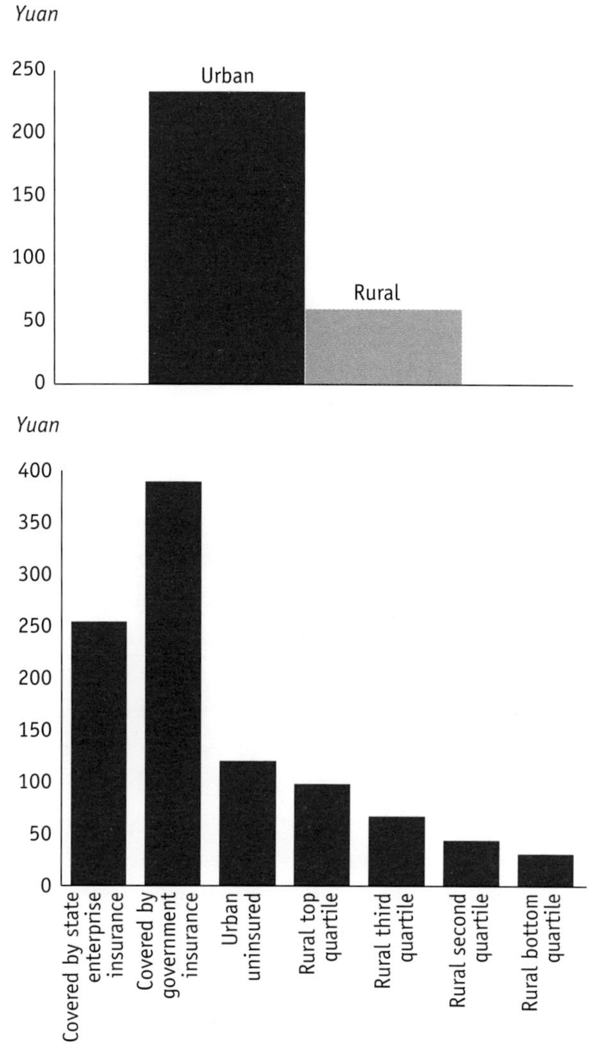

Source: Annex table A.6; Wei 1995.

Those covered receive free care in the government's clinics and hospitals or are reimbursed by the government agency that employs them. Beneficiaries include employees and retirees of central, provincial, and local governments, disabled veterans, and university students.

There is a discrepancy between official estimates of the coverage of the government insurance system and the number of people who report that they are covered. In a 1993 health survey 5.8 percent of those interviewed—equivalent to 70 million people—said that they were covered by the system. The difference may be due to dependents stating that they are eligible even though the government may not count them as eligible. About 10 percent of those who said they were covered by the system in 1993 lived in rural areas. More analysis of eligibility and coverage is needed, especially in light of the very high per capita spending revealed by official estimates of the cost of the insurance program (Zhao Yuxin 1995).

The group with the second-best health insurance coverage is the 11.7 percent of the population employed by state enterprises. The state enterprise insurance system originated in 1951, when the government adopted a policy requiring enterprises to provide or finance health services for employees and retirees and to cover 50 percent of the health care costs of dependents. The government requires state enterprises to contribute an amount equal to 14 percent of their wage bill to cover health benefits and such welfare benefits as child care. Eroding profits in recent years have led many enterprises to restrict employee eligibility for insurance-paid health care. Surveys in 1992 and 1993 show that many workers who were in principle covered did not receive any insurance-paid care (Henderson and others 1995).

State enterprises spent 34.4 billion yuan for health care in 1993—more than a quarter of total health spending in China (see table 2.1). A sixth went to their own services, and much of the rest to public hospitals. Enterprises' spending on their own services is about half the size of the government's recurrent health budget. That underscores the risk the government would face in trying to replace the services now delivered directly by enterprises to their workers and retirees. The government strategy for reforming state enterprises recommends separating the health services they own and manage from their principal business. The health insurance centers created as part of demonstration projects in Jiujiang and Zhenjiang in late 1994 may offer a model for doing so (see chapter 7).

Health coverage in rural areas

In the 1960s and 1970s the rural cooperative medical system reached most rural Chinese. Under this system village authorities used funds from agricultural collectives to pay for the training and salaries of barefoot doctors, locally recruited health workers who met villagers' basic health needs. The collectives helped to pay for farmers' health care, but most services required substantial copayments. The system was not without problems: many village cooperative medical systems had only enough revenue to function from autumn to spring. By 1985 fewer than 10 percent of China's villages maintained cooperative arrangements, and most of these were villages ranking among the top fourth in income (Zhao Zhuyan and Lusheng Wang 1995). (The decline of the cooperative medical system is discussed in more detail in chapter 6.)

Some vestiges of the system remain even in poor villages, however. Village doctors throughout China sell contract or prepayment insurance for immunization and maternal and child health care services. One-year and four-year contracts are available, depending on how long the series of shots and dosages will take to administer. A typical immunization contract provides four years of immunizations—the Expanded Program on Immunization standard groups of DPT (diphtheria, pertussis, and tetanus), measles, and polio—at a prepaid price of a few yuan. The maternal and child health care contract covers antenatal and postnatal care for a child up to age seven and includes nutrition, growth monitoring, and referral, if necessary. These insurance contracts reach 40 percent or more of the children in the local area.

Community financing schemes similar to the former cooperative medical system still operate in the rural areas of some wealthy provinces, such as Jiangsu. These schemes use community health funds derived from township enterprises, village tax revenue, and voluntary contributions to pay for most medical fees and pharmaceuticals. A local township health center committee supervises the use of the funds. Each village clinic has its own account to pay for drugs and other materials. The salaries of the village doctors depend on the number of patients served, not on pharmaceutical sales.

For the village clinics in Shanxi Province, a different financing pattern prevails. They rely on community financing based on an annual prepaid fee (4 yuan per person) covering four services (pharmaceuticals are excluded). The villages provide the clinics with office space, and village doctors make their income from selling drugs: the more they sell, the higher their income. In October 1994 most of the village clinics in rural Jinzhong prefecture, Shanxi, resembled drugstores. Those visited by one researcher had stocks of more than 1,000 kinds of drugs (Jin 1995a).

Trends in how people pay for health care

Most Chinese—some 800 million in rural areas and perhaps 100 million in urban areas—pay directly for health services when they receive them. That can have implications for whether people seek care. A 1992–93 survey found that of those who had been referred to a hospital for care, 40.6 percent did not seek hospitalization on grounds of excessive cost and inability to pay (Zhao Zhuyan and Lusheng Wang 1995). Even middle-income farmers are unlikely to have enough savings to pay for a long hospital stay. The share of out-of-pocket health spending has risen steadily since the late 1970s. Thus while many countries are moving toward a curative health care system that is financed publicly but provided largely privately, China is moving in the opposite direction.

Notes

This chapter updates earlier World Bank health sector reports on China (1984 and 1992a).

1. In January 1996 the Health Economics Institute in Beijing revised its estimate of national health spending in 1993, raising it from 3.6 percent to 3.8 percent of GDP. It has not revised estimates for earlier years.

chapter three

Strengthening Public Health Programs

Public health programs address the health problems of entire populations or population groups. They may provide a specific health service for the community (immunizations), promote healthy behavior (reducing tobacco consumption, limiting salt intake, avoiding sexually transmitted diseases), improve the safety of the environment (sanitation), or detect and monitor the incidence of disease. Clinical services for the treatment of certain infectious diseases—such as tuberculosis and sexually transmitted diseases—are also considered public health activities.

Most public health programs thus provide services that yield important social benefits, but for which individuals are unwilling to pay the full cost. Governments therefore have a large role to play in ensuring the provision of public health services. Although China has a long history of well-developed public health programs producing impressive results, this role needs to be strengthened in China today.

China's public health programs have addressed tropical diseases (malaria, schistosomiasis), micronutrient deficiencies, sanitation, and sexually transmitted diseases and have promoted breastfeeding. But in recent years its programs have faced funding difficulties and coordination and other operational problems. In addition, China's disease patterns and health risks are changing, and public health programs need to adapt to new challenges.

This chapter reviews the status of public health programs in China today. It is not intended to be exhaustive, but instead uses immunizations, tuberculosis control, and antitobacco efforts to illustrate problems in current public health programs and suggest broad directions for change. It concludes that public health programs in China need much strengthening—in public finance, program strategies, and content.

The weakening structure and finance of public health programs

The Ministry of Health, under the authority of the State Council, provides technical leadership and sets guidelines on public health activities as part of its leadership of the health sector. The Epidemic Prevention Service is the backbone of public health programs in China. Many other agencies also carry out public health activities or public health research, including the General Office of the National Patriotic Health Campaign Commission and its local branches, the Chinese Academy of Preventive Medicine, the Center for Health Statistics and Information, and the National Institute for Health Education.

The Epidemic Prevention Service employs a quarter of a million workers and extends disease control programs throughout rural China through its county-level epidemic prevention stations. It maintains the cold chain (refrigeration equipment) for immunizations, makes field visits to ensure water quality, is responsible for the control of diarrheal diseases, and runs endemic disease control programs in many areas (such as those for malaria and schistosomiasis). For several decades the agency was fully funded, for both staff and operating costs, from provincial budget outlays that drew on general revenues. Until the 1980s public health activities were carried out in a highly organized fashion under the supervision of the Epidemic Prevention Service, which drew on village doctors, township health centers, and county hospitals as needed.

China's fiscal decentralization in the early 1980s weakened both the financing and the coordination of public health activities. It gave much more budgetary autonomy to local governments, and provincial health bureaus now develop their own programs according to national guidelines. County hospitals, epidemic prevention stations, maternal and child health centers, and township health centers continue to receive some public subsidies for salaries, but they are now required to generate substantial revenue from user fees. The fiscal decentralization means that poorer counties now have the least capacity to develop and maintain public health programs. It has also weakened coordination among the epidemic prevention stations, township health centers, and village doctors. The epidemic prevention stations have difficulty supervising and influencing the activities of village doctors, who now operate as independent practitioners and generate income from fee-for-service medicine.

Public financing of the Epidemic Prevention Service has remained at about 1.5 billion yuan since 1986. But as a share of GDP it fell from 0.11 percent in 1978 to 0.04 percent in 1993 (table 3.1). The agency faces funding difficulties because of the rising cost of inputs and the increasing budget needed to support retired health workers. To cover the costs of its services, it has had to rely increasingly on its own revenues. In 1993 it generated an estimated 1.6 billion yuan in revenue from fees paid by individuals and institutions, more than the 1.53 billion yuan it received from the government budget.

TABLE 3.1
Financing of the Epidemic Prevention Service in selected years, 1978–93

	Public financing			Revenue from user fees
Year	Total (billions of 1993 yuan)[a]	Per capita (yuan)	As a share of GDP (percent)	(billions of 1993 yuan)
1978	0.94	1.0	0.11	—
1982	1.13	1.1	0.10	—
1986	1.54	1.4	0.09	—
1990	1.58	1.4	0.06	—
1993	1.53	1.3	0.04	1.6

— Not available.
a. These numbers differ from those in table 2.1 because they include both the recurrent and the capital budget.
Source: Wei 1995 and 1996.

The increasing reliance on user fees has had costs. Over the past several years ancillary services whose costs cannot be recouped through user fees have been cut. For example, in Shuoyang County, in Shanxi Province, 80 percent of the Epidemic Prevention Service budget went to staff salaries in 1993, while the number of days assigned to fieldwork in villages in 1994 fell to less than a quarter of what it had been five years earlier. Some staff have shifted their attention to the services for which fees can most easily be charged, such as food inspections, although these services are not necessarily the highest priority. And fees have reduced demand for such services as tuberculosis control and preventive health services, particularly among the poor.

Immunizations—improve coordination and funding

The Expanded Program on Immunization is a key public health program in China. Since the late 1970s China's government has provided political and financial support for immunization against tuberculosis, diphtheria, pertussis, tetanus, polio, and, more recently, measles. Coverage of the four basic vaccines (DPT, polio, measles, and BCG, the vaccine to prevent tuberculosis) reached at least 80 percent in all provinces in 1988 and nearly 80 percent in all counties by 1990—with dramatic benefits. The incidence of pertussis decreased from 126 to 1 per 100,000 between 1978 and 1993, and that of measles from 250 to 10 per 100,000.

China recently expanded its immunization goals. The government hopes to eradicate polio from China soon, and it added hepatitis B to the Expanded Program on Immunization, although it is not yet clear how broad the coverage of newborn infants will be. Given the high levels of postnatal transmission in China, immunization at birth against hepatitis B is a cost-effective way to prevent liver cancer and cirrhosis.

But the immunization program faces major challenges. With the decline in funding for the Epidemic Prevention Service, transport and refrigeration facilities are breaking down, and many local programs face shortages in supplies. Major system breakdowns occurred in 1993 and 1994 in many poor areas, leading to a decline in immunization coverage. Village doctors, now operating as independent practitioners, no longer coordinate closely with the Epidemic Prevention Service—although they carry out immunizations for a flat fee with vaccines it provides. But the fee does not cover their costs (for needles, syringes, sterilization, and operation of the cold chain), which can lead to inappropriate sterilization practices and loss of vaccine potency. The Epidemic Prevention Service and the village doctors charge for immunizations in many areas, raising a financial obstacle for poor households. Moreover, to generate additional revenue, some program staff perform unnecessary antibody tests before providing immunizations. Finally, coverage remains uneven (table 3.2). Measles coverage in rural Shanxi and Guizhou is as low as that in many Sub-Saharan African countries. And in the cities the growing numbers of unregistered residents—the "floating" population—lack access to these public health services.

Additional funding is needed to consolidate and expand China's gains in immunization. In this and other priority public health programs the government needs to ensure adequate salaries and supplies, appropriate training and supervision, performance-related incentives—for both Epidemic Prevention Service staff and the village doctors who coordinate with them—and good working conditions to maintain the commitment of public health workers.

Tuberculosis control—expand and subsidize the new program

Tuberculosis control in China illustrates both what can be achieved with a well-run public health program and what can go wrong. Although the death rate from tuberculosis is decreasing, the disease remains a major health problem in China, accounting for an estimated 3 percent of deaths in 1990 (Murray and Lopez 1996). Active tuberculosis is a highly infectious disease, and

TABLE 3.2
Immunization coverage in China, 1993
(percent)

Area	BCG	Polio	DPT	Measles
National	90	88	86	85
Urban	95	94	93	85
Rural	85	86	84	73
Shanxi Province	65	66	62	49
Guizhou Province	72	74	67	53

Note: National, urban, and rural figures for measles are inconsistent.
Source: World Bank 1995a, p. 3.

public health programs throughout the world try to identify and treat infected people early in the course of the disease to prevent transmission to others. Without appropriate treatment, 60 percent of tuberculosis patients will die. Well-run programs can cure 80–90 percent of patients, poorly administered programs 30 percent or fewer. To avoid financial barriers to tuberculosis treatment, most programs provide treatment free of charge, and some even pay patients to comply with the treatment. China made much progress in controlling tuberculosis during the 1960s and 1970s using standard antibiotic treatment that was essentially free of charge.

But the changes in China's health financing in the 1980s diminished the effectiveness of tuberculosis control programs. As public subsidies were reduced, public facilities were encouraged to recoup their expenses from user fees. That led to many distortions. When doctors and hospitals expected to be reimbursed by the government or state enterprise insurance system, they performed excessive diagnostic tests and examinations during treatment and dispensed higher-cost antibiotics that should have been reserved for more resistant cases. The cost for drugs plus all other services—including unnecessary laboratory exams, x-rays, and traditional medicines—could total more than 1,000 yuan. Daunted by the cost, many low-income people infected with tuberculosis failed to enter treatment or dropped out early. There were no incentives to ensure that patients completed their treatment. As a result, many cases of tuberculosis remained infectious, and the disease spread to others. And as a direct result of poor treatment practices, the spread of drug-resistant strains has accelerated in China since the 1980s.

Recognizing the problems that result from charging for tuberculosis therapy, China has launched a new tuberculosis control program, already operating in some areas. In line with recommendations of the World Health Organization, the program emphasizes directly observed short-course chemotherapy, subsidies for treatment, and appropriate incentives for care providers. Health providers refer patients with symptoms suggestive of tuberculosis to the tuberculosis county dispensary under the Epidemic Prevention Service for physical examination and fluoroscopy. Patients who test positive for tuberculosis are treated with short-course chemotherapy, with every dose observed by the village doctor. Drugs are provided free of charge. The village doctor receives an initial payment when the patient is enrolled, another payment after two months, and a final payment after treatment is completed. Tuberculosis cases are closely monitored at the county level.

The program emphasizes supervision. At the beginning of treatment, county tuberculosis dispensary staff and township disease control officers meet each patient and the village doctor managing the patient to review the treatment plan. Other aspects of the program are also supervised, including the laboratory protocol and county registry.

In the first four years almost 1.6 million patients suspected to have tuberculosis were referred to the program. The cure rate under the program is 90 percent among new cases, compared with about 50 percent before. The failure rate in previously treated cases, which is an indicator of drug resistance, fell from 18 percent in 1991 to 6 percent in early 1994 (China Tuberculosis Control Collaboration 1996). This experience shows that careful supervision, adequate funding, and appropriate financial incentives for providers can make a dramatic difference in addressing a major public health problem. Unfortunately, many patients are still charged for treatment (those outside the project area and those not referred to the tuberculosis program for treatment) and their cure rates are low.

Antitobacco efforts—a two-pronged approach

Smoking is a major health problem in China. If current smoking patterns persist, about 50 million Chinese now aged 0–19 will eventually die as a result of smoking (Peto 1986). According to a 1984 nationwide survey, 61 percent of men and 7 percent of women in China smoked manufactured cigarettes (table 3.3). More recent surveys indicate that smoking is becoming even more widespread. Men are far more likely than women to smoke—Chinese men make up about 10 percent of the adults in the world, but consume 30 percent of the world's cigarettes. And the lower a person's education level, the more likely that he or she is a regular smoker.

More than 800,000 deaths in China in 1990 were attributable to smoking, including deaths from coronary disease, chronic obstructive lung disease, and lung can-

TABLE 3.3

Share of Chinese who smoke, by gender, age, occupation, and education, 1984
(percent)

Group	Male		Female	
	Regular smokers	Occasional smokers	Regular smokers	Occasional smokers
All	55.5	5.5	6.4	0.7
Age (years)				
20	14.1	5.0	0.2	0.1
20–60	63.3	6.0	5.8	0.6
Over 60	63.6	3.0	16.5	1.4
Occupation				
Worker	60.7	5.0	6.6	0.7
Peasant	58.1	5.7	6.0	0.6
Cadre	54.9	4.3	5.6	1.0
Teacher	44.8	5.3	2.6	0.3
Doctor	49.1	7.6	2.2	0.2
Education				
College	39.6	5.2	2.8	0.8
Middle school	47.5	6.3	1.7	0.2
Primary	61.2	5.3	5.5	0.6
Illiterate	63.9	4.0	10.6	1.0

Source: Data from 1984 National Survey on Smoking in China, as cited in Teh-Wei Hu 1995 and China, Ministry of Health 1991.

cer (Murray and Lopez 1996). Present patterns of smoking suggest that this number will increase dramatically over the next thirty years to at least 2 million deaths a year. The economic losses in terms of health care costs are also huge: the health spending related to smoking was estimated at 6.9 billion yuan in 1989 (Jin 1995d).

Effective public health programs to reduce tobacco use are based on two complementary strategies: taxing tobacco, to provide a financial deterrent to consumption, and influencing behavior through such measures as health education, banning smoking in public places, and banning cigarette advertising and promotion. Prohibiting cigarette advertising and promotion in China, particularly by foreign tobacco companies, could be effective not only in addressing the high prevalence of smoking among men, but also in preventing an increase in smoking rates among women.

China already has a cigarette tax in place. In fact, the cigarette tax is a major source of government revenue, accounting for 31 billion yuan, or 9.5 percent, in 1992 (Teh-Wei Hu 1995). The effective tax rate on cigarettes was about 38 percent in 1991, compared with 60 percent in Japan and the Republic of Korea and 85 percent in Denmark. Several Chinese cities have banned cigarette advertising and smoking in public places. But tobacco companies are circumventing these restrictions through sports sponsorship and other forms of promotion, and more action is needed. The government is considering increasing the cigarette tax and using the revenue to finance antitobacco campaigns and other public health efforts.

The amount of revenue that an increase in the tobacco tax would generate depends on the price elasticity of demand in China. In other countries the price elasticity has been estimated at between –.35 and –.74 (table 3.4). Assuming a rough average elasticity of –.5 in China, a 10 percent increase in tobacco prices would reduce demand by about 5 percent and generate an additional 5 billion yuan of revenue a year. A 20 percent incremental tobacco tax would generate an additional 10 billion yuan in revenue.

An ad valorem tax (a tax based on a fixed proportion of the retail price) is preferable. The tax would adjust with changes in tobacco prices. And it would also be more progressive than a flat rate, because high-income smokers who purchase more expensive brands would be willing to pay more. Also preferable is to impose the tax on all tobacco products, to reduce the possibility of substitution.

A 10 or 20 percent incremental tax on tobacco products—accompanied by increased health education, bans on tobacco advertising and sports sponsorships, limits on the tar and nicotine content of manufactured cigarettes, and media campaigns against tobacco—would generate important health benefits in China. Because the rationale for raising the tobacco tax is to reduce tobacco-related illness now and in the future, there is an argument for using all or part of the revenue from the

TABLE 3.4

Price elasticity of demand for tobacco in selected countries, various years

Country	Elasticity
Finland	–.35
Ireland	–.38
United Kingdom	–.39
United States	–.50
Switzerland	–.50
Austria	–.54
Canada	–.74
Median	–.45

Source: Zimring and Nelson 1995.

tax for health education programs and other antismoking activities. The additional revenue could also be used to support other priority public health programs.

Conclusion

China needs a strong, publicly funded public health service that can adapt to the changing risk factors and disease burden in the country. This kind of public health service could be achieved by a restructured and reformed Epidemic Prevention Service working in coordination with other agencies. Since public health services must be financed by the government if they are to be provided at socially optimal levels, central and provincial governments should fully fund those with high priority. Central and provincial funding is also necessary to ensure that the poorest counties, which have some of the biggest public health problems, can pay for their programs.

Funding for activities that are now the responsibility of the Epidemic Prevention Service needs to be increased substantially in the medium term. To fully fund tuberculosis control, the Expanded Program on Immunization (including universal hepatitis B immunization for infants), endemic disease control, health education, and other important programs would require raising funding from 1.3 billion yuan in 1993 to at least 6.5 billion yuan in 2001. The new tuberculosis program needs to be extended throughout the country, with treatment provided free of charge. And special outreach measures are needed for the unregistered urban population.

BOX 3.1

The challenges of STDs and HIV in China

Sexually transmitted diseases (STDs) and the human immunodeficiency virus (HIV), the virus that causes AIDS, pose important public health challenges in China today. The World Health Organization estimates that China had 100,000 cases of HIV infection at the end of 1995. This is a relatively low number, but China faces the risk of rapid spread of HIV.

Several factors contribute to this risk. With economic modernization, the increasing mobility of the population can speed the spread of the disease. The rapidly growing migrant population is another problem: migrants tend to have little access to medical services and little knowledge about HIV/AIDS, and migrant women from rural areas are more likely than others to engage in commercial sex work. Many vulnerable groups have almost no knowledge about HIV and how to prevent its transmission. Finally, the blood supply in China is poorly screened and overly commercialized.

The areas where HIV is most prevalent in China and the modes of transmission are changing. Until recently, about 80 percent of the reported cases of HIV infection were in the southwestern province of Yunnan, where drug trafficking and prostitution are major problems. But the number of cases in coastal areas and large cities is now rising. While about 64 percent of current HIV infections in China result from the use of injected drugs (China, Ministry of Health 1995b), the government believes that the main mode of transmission in the coming years will be sex. High-risk sexual behavior is increasing, as evidenced by China's rising STD rates. STDs have reemerged as a significant public health problem in China, after major efforts to eradicate them in the 1950s and 1960s. Reported STD cases rose from 5,838 in 1985 to 237,573 in 1993 (China, Ministry of Health 1995b).

Concerted prevention efforts now, while the prevalence of HIV is still relatively low in China, would have a huge payoff. The cost-effectiveness of interventions drops sharply when infections cross from high-risk groups to the general population. Moreover, increased prevalence of HIV is likely to generate many other health problems. It would greatly exacerbate the already large problem of tuberculosis, for example, one of the major opportunistic infections of HIV. Prevention efforts should include disseminating information on how to avoid infection, promoting condom use, treating other sexually transmitted diseases, and reducing blood-borne transmission. A growing body of research shows that preventing and treating STDs has a significant effect in averting HIV transmission.

To be effective, efforts to control the spread of HIV and STDs must be multisectoral. The government's National Strategic Plan for the Prevention of AIDS and STDs lays out a plan that would involve twenty-two government agencies and groups in society in prevention activities aimed at such groups as drug users, prostitutes and their clients, STD patients, migrants, transport workers, overseas laborers and businessmen, and men who have sex with men (China, Ministry of Health 1995b). The Ministry of Railroads, for example, could help in promoting condoms and in educating migrant groups about AIDS.

With the support of the World Bank–financed Disease Prevention Project, the Chinese government has recently embarked on programs to help prevent and control the spread of STDs and HIV in seven major cities and in Yunnan Province. These programs will strengthen central planning and implementation capacity. They focus on intersectoral coordination, policy reforms (for example, to promote anonymity of treatment, improved availability and use of condoms, and sex education in schools), behavior risk factor surveillance, training of health workers, improved management of STD cases, and other health promotion measures.

To make effective use of the increased resources, the government needs to ensure that public health programs are implemented efficiently. It needs to provide for adequate salaries, appropriate training, performance-related incentives, and good working conditions to maintain the commitment of public health workers. It needs to finance and develop new disease control programs that address the changing pattern of disease in China. And it needs to ensure that China's highly effective health surveillance system is maintained and adapted to the changing disease burden and risk factors. Intersectoral cooperation and collaboration will become increasingly important, as many public health problems require solutions beyond the scope of the health sector alone, such as lead in the air, indoor air pollution, drug abuse, traffic accidents, suicides, and HIV/AIDS (box 3.1).

Another recommended public health action is a substantial increase in the tobacco tax, accompanied by the strengthening of other measures to reduce smoking (see chapter 8). These actions would help reduce costly illness and premature death related to smoking, and the additional revenue could help finance public health programs (as in Australia and California).

chapter four

Meeting the Needs of the Poor

Poverty is both an important cause and an important consequence of ill health. The poor are more likely to suffer from ill health, and their health problems can keep them in poverty. Thus there is strong justification—on grounds of reducing poverty and improving equity—for government subsidies to improve the access of the poor to important health services. The Chinese government now provides minimal subsidies for this purpose. Resources need to be redirected or expanded to ensure that key services reach the poor. And because public resources are scarce, subsidies need to be carefully targeted to the very poor and to the services that provide the greatest impact on health for the money spent.

China has experienced an enormous reduction in poverty since economic reforms began. But progress has been unsteady. Strong rural growth drove rapid improvements in the quality of life for the poor in the early 1980s. The pace then stalled as the locus of economic growth shifted to urban and coastal regions. Most of the poor in China are now concentrated in resource-deficient rural areas. In upland sections of the interior provinces of northern, northwestern, and southwestern China entire communities are poor. The health status in these areas of concentrated poverty is bleak. Infant and maternal mortality rates in very poor counties are at least 50–100 percent greater than the national average, and in the poorest townships and villages much greater yet.

There are signs that poverty again began to fall in the early 1990s. While the trends are clear, the number of Chinese considered poor depends on the poverty line. The government's definition of poverty yields some 80 million–100 million poor (table 4.1). But when international norms adopted by the World Bank for minimally acceptable living standards ($1 a person per day) are used, the figure for 1993 jumps to 350 million (World Bank 1996c).

Absolute poverty is largely a rural problem in China. But there are also large numbers of urban poor, and these people too have little access to public health programs and clinical services, particularly the unregistered migrant populations.

A broad range of measures is needed to reduce poverty in China, including policies to improve labor mobility and foster rural enterprise and agricultural development in poor areas. Targeted subsidies to improve the health status of the poor would also improve their welfare and economic productivity. This chapter discusses how targeted health assistance might best be achieved and the likely costs of such assistance. It focuses on the poor in rural areas, where absolute poverty is most prevalent in China.

TABLE 4.1

Incidence of absolute poverty in China in selected years, 1978–90

Population group	1978	1985	1990
All			
Number of poor (millions)	270	97	98
Incidence of poverty (percent)	28.0	9.2	8.6
Urban			
Number of poor (millions)	10	1	1
Incidence of poverty (percent)	4.4	0.4	0.4
Rural			
Number of poor (millions)	260	96	97
Incidence of poverty (percent)	33.0	11.9	11.5

Note: By convention, the Chinese statistical system treats unregistered urban residents as rural residents, so the figures for urban areas are underestimates. The poverty line used in this table derives its initial monetary value from a 2,150-calorie-a-day food consumption basket deemed to supply adequate nutrients for normal human functioning. It then adds a sum for expenditures on nonfood commodities and services based on the average expenditure pattern of the poor.
Source: World Bank 1992b.

Health problems of the poor

The poor face a greater disease burden than the nonpoor. According to the World Bank's *World Development Report 1993,* the poor are much more likely than the better-off to suffer from certain childhood diseases, tuberculosis, maternal health problems, micronutrient deficiencies, and sexually transmitted diseases (World Bank 1993b).

The rural poor

This pattern holds true in China. Much of China's infectious and parasitic disease burden, including tuberculosis, diarrheal disease, and iodine deficiency disorders, is concentrated in poor and remote areas. Roughly half of children in households at or below the absolute poverty line are at least mildly malnourished (stunted), and iron, vitamin A, and other micronutrient deficiencies remain common among the poor. As many as 90 percent of poor children suffer chronic worm infections. The poorest quartile of the rural population reports an infectious disease rate three times that of the richest quartile of the rural population, and an infant mortality rate more than twice that of the richest quartile (table 4.2).

The rural poor not only have a greater disease burden than higher-income rural residents; they use health services less. A third of low-income households seek no health care, according to the Study of Thirty Poor

TABLE 4.2

Health status by income quartile in rural China, 1993

Indicator	Top quartile	Second quartile	Third quartile	Bottom quartile
Average per capita income (yuan per year)	927	677	561	441
Incidence of infectious disease (per 1,000)	3.3	5.1	5.4	9.5
Infant mortality rate (deaths per 1,000 live births)	29	34	44	72
Life expectancy (years)	71	69	68	64

Source: The 1993 National Health Services Survey and the 1990 population census.

Counties, compared with only 16 percent of high-income rural households (China Network and Harvard School of Public Health 1996). On average, people in the lowest income quartile make only 60 percent as many health care visits as those in the highest quartile, and spend only 50 percent as many days as a hospital inpatient. Data from the Study of Thirty Poor Counties also show that the poor spend a larger share of their cash income on medical care: 23 percent for those with household cash incomes under 250 yuan a month, compared with 11 percent for those with incomes between 430 and 690 yuan a month (Luo Wujin as cited in Hammer 1996).

The urban poor

Much less is known about the health problems of the urban poor, particularly the unregistered urban poor. Compared with other urban residents, the urban migrant population is much more likely to live in crowded, unsanitary conditions and faces higher risks of contracting communicable diseases such as tuberculosis. China's cities have few community-based health service facilities, and the outreach of public health programs to the unregistered urban population is weak at best.

The role of government spending in health care for the poor

Government spending on health has not been effective in reaching China's poorest residents. China's 1981 public finance reforms devolved public finance to the provincial and county level, limiting the central government's ability to redistribute funds from richer to poorer areas of the country. An analysis of public expenditure over eleven years shows that the allocation of public expenditure is skewed toward richer regions and, within regions, to the provinces growing fastest (Hammer 1996). Within provinces, government spending is concentrated on government health insurance and hospital inpatient and outpatient care. Services that disproportionately benefit the poor, such as the Maternal and Child Health Program and the Epidemic Prevention Service, have been constrained and increasingly forced to rely on revenue from user fees. Not surprising, traditional public health activities achieve the greatest coverage in the wealthiest provinces. For example, while more than 90 percent of the children in China's richest provinces receive their third dose of DPT, less than half the children do in the poorest provinces (Hammer 1996).

Bringing key health services to the rural poor

The State Council Leading Group Office of Poverty Alleviation and Development has identified 592 counties with the greatest incidence of rural poverty (table 4.3). These counties, which make up a quarter of all Chinese counties, have a total population of about 210 million. In 1992–93, 58 percent of the population in these counties had annual incomes below 500 yuan and about 26 percent had incomes below 300 yuan. Experts disagree on how many of China's absolute poor live in the 592 poor counties, but generally agree that at least

TABLE 4.3
Population with per capita income below selected cutoff lines in China's 592 poor counties, 1992–93
(percent)

Province	500 yuan	300 yuan	200 yuan
Yunnan	83	52	20
Guizhou	75	48	23
Zhejiang	74	26	..
Sichuan	74	29	7
Gansu	68	27	10
Shaanxi	66	29	6
Ningxia	64	34	9
Hebei	64	24	10
Shanxi	63	16	4
Hunan	60	29	11
Guangxi	58	35	11
Henan	58	22	6
Hubei	57	23	3
Qinghai	51	21	6
Xinjiang	42	19	8
Anhui	38	9	1
Nei Menggu	37	15	2
Jilin	33	20	12
Hainan	30	14	3
Guangdong	26	2	..
Jiangxi	23	5	1
Shandong	22	4	2
Liaoning	17	4	..
Heilongjiang	15	8	3
Fujian	6	1	..
Xizang	—	—	—
All	58	26	9
Total (millions)	121.4	55	18.4

— Not available.
.. Less than 0.5 percent.
Source: Data provided by Chinese authorities.

half do. These counties are therefore a useful reference point for a discussion of subsidies to improve the poor's health.

The key issue is targeting: how to reach the poor (and only the poor) with the appropriate transfers without incurring excessive administrative or political costs. There are several ways to target the poor:

- Geographical targeting to areas where the poor are concentrated. For example, poor villages in China's 592 officially designated poor counties could be targeted for subsidized services.
- Individual or household targeting, by identifying the poor and certifying their eligibility for subsidized services. (China may be the only developing country where this is feasible because of good government records, but this kind of targeting is administratively costly.)
- Program targeting to health services that particularly benefit the poor, such as deworming and management of acute infections in children. (Such services would be in addition to the public health programs discussed in the previous chapter.)

This report recommends a blend of geographical targeting (probably most practical at the village level) in China's poor counties and program targeting for a few services that particularly benefit the poor. Program targeting could cover the entire population in China's poor counties (providing an element of geographical targeting) or the nation. Targeting at the national level would be more expensive, but it could ensure that services reach the one-third of the absolute poor who do not live in officially designated poor counties. The financing for subsidies would have to come largely from the central and provincial governments, since poor counties have little means to finance such programs.

Under a program of geographical targeting, the government might direct subsidies to health care providers in poor villages (supply-side subsidies). Providers would then offer services free or at low cost to village residents. The government could determine what services would be provided—for example, choosing highly cost-effective services to ensure maximum health impact from the program. If a village was part of a viable community financing scheme, the government could channel the subsidies to the scheme (demand-side subsidies) rather than to the health care providers (see chapter 6).

Under program targeting, the government might fully fund health interventions addressing conditions that disproportionately affect the poor. Eleven major disease conditions can be identified in China as almost entirely associated with poverty (table 4.4). Together they accounted for an estimated 23 percent of China's disease burden in 1990. These conditions can be largely addressed by cost-effective public health and clinical interventions, such as immunization, short-course chemotherapy (for tuberculosis), iodized salt treatments, deworming, and prenatal and delivery care.

TABLE 4.4
Poverty-related disease conditions in China, 1990
(percent)

Condition	Available cost-effective intervention[a]	Share of total disease burden
Infectious and parasitic diseases		
Tuberculosis	Short-course chemotherapy	2.0
Diarrheal diseases	Monitored oral rehydration	1.8
Measles, polio, pertussis, diphtheria, tetanus	Immunization	1.1
Intestinal nematode infections and anemia	School-based deworming	0.7
Other infectious and parasitic diseases	Acute chemotherapy	1.2
Lower respiratory infections	Early identification and antibiotic treatment	5.7
Maternal conditions	Emergency obstetric services with prenatal care	1.2
Perinatal conditions	Obstetric care from trained personnel	4.9
Nutritional deficiencies		
Protein-energy malnutrition	Breastfeeding, prevention and treatment of childhood infections	1.0
Vitamin A and iodine deficiency	Supplementation, food fortification	0.4
Anemia	Supplementation, food fortification	3.2
Total		23.1

Note: The table highlights only the main disease conditions that are almost entirely conditions of the poor. The poor also suffer disproportionately from many others, such as diabetes, chronic obstructive pulmonary disease, sexually transmitted diseases, and HIV infection.
a. Interventions costing less than $150 per disability-adjusted life year.
Source: Murray and Lopez 1996; World Bank 1993b.

Per capita health spending, public and private, now stands at about 60 yuan a year in rural areas. The public subsidy for a package of services for poor villages in poor counties might be say, 25 yuan per capita to finance a 30 yuan package (with the other 5 yuan from other sources). The services that could be provided under a 30 yuan-per-person package are described in box 6.1. With income growth projected to double between now and 2001, the per capita costs of this package could be assumed to double as well. If the subsidy was targeted to the population of about 75 million in poor villages in poor counties, the cost would be about 3.75 billion yuan in 2001. An additional 1 billion yuan or more might be directed to program subsidies, either for the entire population in poor counties (210 million) or nationwide. These cost estimates would differ with different assumptions about the population to be targeted and the size of the per capita subsidy in 2001.

The subsidies should be accompanied by systematic monitoring and evaluation to determine whether the programs indeed help the poor and to guide policy improvements.

Bringing key health services to the urban poor

Although poverty is concentrated in rural areas, measures are also needed to improve access to essential health services for the urban poor, who are not covered by the government and state enterprise insurance schemes. An important first step would be to fully fund priority public health programs with public finance (see chapter 3). To improve access to clinical care, most of which in urban areas is provided by hospitals, China could consider measures to revitalize community care, especially in poor neighborhoods. Some of the subsidies that now go to hospitals could be gradually redirected to services for the urban poor at community clinics or hospitals. If some urban areas form a community with sufficient stability and cohesiveness to enter into a social contract, community financing schemes might be a feasible channel for financing and organizing services for the urban poor (Hsiao 1995b). Public subsidies for interventions that particularly benefit the poor could also help low-income urban residents.

chapter five

Reforming Pricing and Planning

Government policies can play a big part in determining the efficiency of the health sector—whether health spending is allocated to the "best buys," whether services are produced at least cost, and whether care is clinically appropriate. The key is well-designed incentives.

The present pricing structure in China's health sector provides incentives for excessive and inappropriate care, producing major distortions and inefficiencies. The parallel public delivery systems are another major source of inefficiency. The excess capacity and idle resources in these systems represent much waste.

This chapter reviews these efficiency issues and proposes reforms in pricing and in the planning and coordination of infrastructure and human resources that could significantly improve the health sector's efficiency in both rural and urban areas.

Ending price distortions in the health sector

Prices for health services in China are set under guidelines established by the Price Commission, often at levels well below costs. Price Commission officials aim to make the prices high enough to protect and develop the services provided, yet low enough to ensure affordability for the users. During the 1960s and 1970s the government tried to increase access to health care by reducing the prices of medical visits and hospital days to levels that a poor farmer could afford. Prices for most medical services in China are still below costs, especially for services with a large labor input. The prices of high-technology diagnostic tests, however, have been set far above costs to offset losses on other services.

The system thus has two pricing extremes. Most services are priced too low, leading to under-the-table payments to physicians and other problems. At the same time, high profit margins on drugs and diagnostic tests encourage overprovision in those areas. For example, now that village doctors' income depends on fees charged for drugs, injections, and diagnostic tests, these doctors have strong incentives to overprescribe drugs and tests. This response drives up health spending without improving health.

At the heart of the issue of pricing policy, as it affects goods and services sold by the health sector to patients, is China's "yellow book" price list, which sets the prices of thousands of medical procedures, services, and diagnostic tests. Despite double-digit inflation in recent years, these listed prices are rarely updated in many provinces, so most health service prices remain fixed at extremely low levels.

The difficulties caused by these pricing policies can be clearly seen in the hospital setting. Most Chinese hospitals derive about 85 percent of their revenues from charges to patients (or the government or state enterprise insurance program they belong to) for each service rendered and drug given. Faced with the irrational price structure set by the yellow book, hospital managers use profitable products to cross-subsidize underpriced products. But to generate a small profit, hospitals must oversell a high volume of profitable services. This "leveraging effect" distorts the patterns of medical services and increases total health costs (box 5.1).

BOX 5.1

The leveraging effect

How much profit hospitals can generate to cross-subsidize services operating at a loss depends on allowed profit margins. For example, drugs have an allowed markup of 15 percent. So, to generate 100 yuan of revenues above cost, a hospital must increase sales of drug prescriptions by 666 yuan. This "leveraging effect" contributes to rapid growth in health spending.

Pricing of health services

The pricing of health services in China has had unintended consequences for the way hospitals make decisions about equipment purchases. The official prices for many routine hospital procedures fall far below the costs of providing those services. In Shanghai the actual costs for some routine procedures, based on modern cost accounting methods, are two to four times the allowed fees for patients paying out of pocket, and two to three times the allowed fees for insured patients (table 5.1). By contrast, the prices of newly introduced diagnostic tests have been set high enough to encourage their rapid adoption. For a CT scan, for example, allowed fees in Shanghai and Tianjin permit a large profit (table 5.2).

As a result of this pricing policy, urban hospitals and even county and township hospitals and health centers have come to see high-technology equipment as their financial salvation (box 5.2). These organizations now routinely organize investor groups to buy such equipment, or they borrow from banks or sell investment shares to staff members. Sometimes hospitals lease equipment from international equipment suppliers,

TABLE 5.1

Production costs and fees for selected procedures in Shanghai hospitals, 1989

(1993 yuan)

		Allowable fees	
Procedure	Average cost	Insured	Self-pay
Appendectomy	118	56	28
Cataract removal	142	44	22
Gastrectomy	326	292	97
Cholecystectomy	199	111	56
Exploratory laparotomy	167	56	28
Hysterectomy	257	139	69
Cesarean section	140	111	56

Source: Chinese Journal of Hospital Management 1993, p. 55.

TABLE 5.2
Production costs and fees for CT scan, 1988
(1993 yuan)

City	Cost			Allowable fees	
	Fixed	Variable	Total	Self-pay	Insured
Shanghai	109	47	156	181	362
Tianjin	80	32	113	181	362

Source: Chen Jie 1994, pp. 4–5.

with the lease payment set at a percentage of the gross revenues generated by the equipment. A hospital's reputation is now said to depend on its possessing the latest equipment, whether computerized tomography, magnetic resonance imaging, fetal monitoring, upgraded intensive care units, burn units, or any of the wide variety of diagnostic techniques on the market. China is in the midst of a diagnostic equipment race among most of its hospitals and many health centers.

Pricing of pharmaceuticals

The price structure for drugs allows markups of 15 percent at both the wholesale and the retail level. Thus hospitals have an incentive not only to overprescribe drugs they purchase from manufacturers, especially expensive drugs, but also to manufacture their own products. They frequently treat patients with intravenous drip solutions of glucose, vitamins, antibiotics, and other drugs. In almost all cases this treatment is provided not because it constitutes justified medical practice, but because it maximizes profit. And village doctors, who earn almost all their health-related income from drug sales and injections, prefer injections over oral prescriptions because they yield higher revenue.

With this incentive structure, it is not surprising that drugs account for 52 percent of health spending in China, compared with an average of 14 percent in OECD countries and 15–40 percent in most other developing countries.[1] Inappropriate use of drugs is not only inefficient, it can also lead to high-cost health problems (for example, the development of antibiotic resistance).

There are several ways to combat overuse and misuse of pharmaceuticals:

- Take the profit out of prescribing. If hospital payments were case-based, they would be independent of the number of drugs and tests prescribed.
- Incorporate high copayments for outpatient drug prescriptions in insurance benefit packages to reduce excess demand.
- Educate providers and consumers. Some countries give providers prescribing handbooks and educate the public on the appropriate use of drugs and the health effects of overuse.
- Develop and enforce hospital formularies and essential drug lists to guide cost-effective prescribing.
- Monitor prescribing patterns and give feedback to physicians. Core indicators of appropriate prescribing patterns include the average number of drugs prescribed per patient encounter, the percentage of drugs prescribed by generic name, the percentage of patient encounters in which an antibiotic is prescribed, the percentage of encounters in which the patient receives one or more injections, and the percentage of drugs prescribed that are not on the relevant essential drug list or in the local formulary for that level of care.

BOX 5.2

The high-technology equipment race

For Jurong County Hospital, in Jiangsu Province, the government subsidy now contributes only a small part of its income: the subsidy declined from 17 percent of revenues in 1985 to only 2.2 percent in 1993. So the hospital has been forced to seek funds from other sources, such as drug sales and tests using high-technology equipment.

In 1993 the hospital bought a CT scanner and other new equipment using capital investments of 0.2 million yuan from the local government, 2.2 million yuan in loans from local banks, and 0.6 million yuan raised from hospital staff. It provided CT diagnostic services to about 1,000 patients in 1994, earning 0.35 million yuan, more than twice its income from high-technology diagnostic testing (mainly CT) in 1993. Although the number of patients grew only 2.1 percent, the number using high-technology equipment increased 50 percent. The hospital's outpatient fees increased 168 percent (Jin 1995a, p. 21).

Such stories are being duplicated throughout China—first in the somewhat richer coastal provinces and, more recently, in the interior. Even Guizhou Province, which ranks lowest in per capita income among China's provinces, has seen an increase in high-technology medicine.

Reforming pricing

Analyses of the Chinese health sector have emphasized the problems caused by pricing policies that distort supply and demand (World Bank 1992a). The standard advice in such situations is to move toward marginal cost pricing, even if fees continue to be fixed. The authorities in China have given much consideration to this policy but have not yet adopted it.[2] Senior policymakers in China seem to accept price reform in principle, but fear that jumps in price could set off negative reactions.

A major study of costs and prices is needed to lay the foundation for reform. Such a study should avoid attempting to revise thousands of prices in detail. Instead, it should focus first on gradually bringing prices closer to costs. Some prices, especially those for services with high labor content, would be raised; others would be lowered. The study should also analyze methods of defining and pricing broad packages of care in line with provider payment reforms (see chapters 6 and 7). And it should devise a mechanism for adjusting prices to inflation, perhaps on an annual basis.

Reallocating government spending

Chinese government spending on health is both inequitable and inefficient. Most is directed to the hospital sector through Ministry of Health hospitals, including township health centers (4.5 billion yuan in 1993) and traditional Chinese medicine facilities (0.8 billion yuan). As discussed earlier, relatively little goes to more cost-effective public health (1.3 billion yuan) and maternal and child health services (0.3 billion yuan). The government insurance system too spends almost all of its money on the hospital sector (11.5 billion yuan in 1993).

Yet government subsidies to hospitals are a small share—no more than 15 percent—of total hospital operating costs. The Ministry of Health has recently argued for more public subsidies to the hospital sector. But increasing the subsidies would only exacerbate the misallocation of resources. Those who benefit most from the present subsidies are middle-income urban residents and those insured under the government and state enterprise insurance systems. The first priority for government resources should be public health and cost-effective care for the poor.

Over the medium term reallocating public spending away from hospitals toward other priorities would be eased by the reforms proposed in this report. Price reform would better align allowable fees with marginal costs, so hospitals would have less need to provide excessive diagnostic tests to cover loss-making activities. Provider payment methods that encourage efficiency could help ensure that hospitals can recover their costs. (Urban insurance experiments are already trying alternative provider payment approaches; see chapter 7.) And greater autonomy in personnel decisions would enable public hospitals to operate more efficiently. Public hospitals in China have considerable autonomy over their budgets, investments, and fee collection, but not their personnel. The government assigns new medical graduates to public hospitals without sufficient regard for staffing needs. As a result, some hospitals have too many doctors and too few nurses, and others have too many personnel in total.

Improving planning and coordination

China has at least three separate vertical systems involved in the planning, financing, and organization of urban hospital facilities—the public health system, the state enterprise system, and the traditional Chinese medicine system. Each vertical system protects its own institutional interests and has little incentive to coordinate with others. Moreover, the public hospitals are owned and managed by different levels of government—national, provincial, and county. In a large city three major hospitals might be located quite close to one another—one operated by the central government, one by the provincial government, and one by a state enterprise. A few kilometers away, across the city line, there could well be a county hospital offering similar services.

In rural areas the collapse of the cooperative medical system weakened the referral and supervision chains between the village, township, and county health organizations. Another problem is the duplication of services between the maternal and child health and family planning facilities and between Western medicine and traditional Chinese medicine facilities, which strains the government's health budgets and leads to misuse of scarce resources (box 5.3). This problem is likely to worsen as the Family Planning Commission extends its facilities to the township level, increasing the duplication.

BOX 5.3

Problems in rural health delivery: A case study of maternal and child health programs

Several health facility and household surveys, a cost-effectiveness study, and a program review by the Ministry of Health have all documented the need to improve maternal and child health services in rural China. Problems in maternal and child health services mirror those facing the rural health system as a whole.

Quality and cost-effectiveness

Because maternal and child health centers at the county level, like many other health providers in China, now must generate much of their income from user fees, they have shifted their focus from preventive public health measures to curative functions for which they can charge fees. The shift to fee services effectively denies health care to part of the rural population, since most rural families no longer have health insurance and many cannot afford fee-for-service care. It has even reduced the coverage of such essential preventive programs as immunization.

The increased emphasis on earning revenue has also led to duplication of services and to quality problems. To generate fees, maternal and child health centers are investing in equipment to perform cesarean sections, for example, rather than referring patients who need this surgery to nearby county hospitals. Such surgical procedures should be concentrated in high-volume centers, where professional skills are more easily kept up to date.

Treatment of childhood diarrhea—still a common problem in some areas of China—illustrates another problem resulting from the need to generate revenue. As hospitals have become increasingly dependent on profits from drugs and services for which public subsidies are highest, such as inpatient care, antidiarrheal treatment in China has come to involve unnecessary hospital stays, intravenous fluids, and potentially dangerous drug treatments. Most cases would be better handled using basic oral rehydration and, when necessary, appropriate antimicrobial treatments at outpatient facilities. The costs of such inefficient service are borne by the patients' families, who pay more than necessary for treatment; the public sector budget; and patients who do not receive treatment because their families cannot afford it.

Organization and coordination

Another problem is the lack of functional coordination among the many different groups providing maternal and child health services. An organization chart of the Ministry of Health shows a number of parallel national programs at the central, provincial, prefectural, and county levels, converging at the township hospital and village level. The Maternal and Child Health Department is generally responsible for maternal and child health policies, although the Medical Administration Department develops hospital care policies, including maternal and child health referral services at the provincial level and below. The Epidemic Prevention Service, under the Department of Disease Control, develops policies on childhood immunizations and diarrheal diseases as well as training sites and materials for managing childhood diarrhea. The Medical Science and Education Department handles health workers' training, including maternal and child health training, although in-service training for maternal and child health is designed and implemented by the Maternal and Child Health Department. This overlap between departments and functions at all levels constrains the development of consistent policies and guidelines, efficient training, and good-quality maternal and child health preventive and curative services.

China needs a rational hospital referral system to make appropriate use of its hospital resources. There appears to be overcrowding in tertiary hospitals, but the average stay of fifteen days for patients in these facilities is excessively long by international standards. With policy changes, many of these patients might be adequately managed at lower levels, where occupancy rates are low in secondary hospitals and township health centers, and the average stay might be shortened.

Regional planning could improve coordination between services. Regional planning could be used to establish referral systems, to develop approaches to disease control, and to guide capital investments in order to avoid duplication and fragmentation of systems. Regional planning efforts should involve all relevant actors, including medical schools, the Ministry of Health, traditional Chinese medicine facilities, and the government and state enterprise insurance systems. As urban insurance centers expand, they would also be a major stakeholder in regional planning efforts. Plans should cover a population base large enough to support comprehensive health services. Experiments with regional health planning in three prefectures show what it can achieve (box 5.4).

Regional planning bodies could be complemented by central and provincial government efforts to better control investments in facilities, expensive equipment, and the mix and number of health personnel. Once expensive and sophisticated facilities are established, they must generate sufficient revenues to fund their operations, raising health care costs in the long term. Many countries have realized too late that they have too many

BOX 5.4

Regional planning for health

Regional health planning in China began under the World Bank–supported Integrated Regional Health Development Project in three prefectures—Baoji in Shaanxi Province, Jiujiang in Jiangxi Province, and Jinhua in Zhejiang Province. In each prefecture a regional health committee made up of government representatives in finance, planning, civil works, and health agreed on a five-year development plan. These experiments in regional planning have had concrete results, attracting increasing attention from senior leaders in China.

The regional plans have led to better disease surveillance and a new tuberculosis control strategy. An antismoking campaign in schools and improved training and outreach for maternal and child health services are in place. Ambulance services in the provincial and regional cities now operate with radio networks. Hospital managers accepted the centralization of high-technology diagnostics in one hospital. Some resources moved from hospitals down to county and township levels. Care for the mentally ill and physical rehabilitation improved. There is better in-service training for medical and health workers, and networks for inspecting and maintaining equipment have been strengthened.

hospital beds or too many physicians or specialists, another source of upward pressure on health costs. And other countries struggle to reform their health systems with too few well-trained hospital managers.

Many countries have found that governments need to play an active part in curbing health investments, both capital and human, because of the long-lasting and powerful effects such investments can have on health service delivery and costs. In the Netherlands, for example, hospitals must apply for approval to purchase specific types of equipment and technology or to provide certain specialized medical services. This regulation has prevented an oversupply of medical technology.

Of course, these policies also have drawbacks. Supply-side controls reduce innovation and restrict market entry of lower-cost providers of specialized medical services. The Netherlands' success has been attributed largely to its severe sanctions: hospitals may be fined, the service may be closed down, or insurance companies may refuse to reimburse hospitals for an unapproved service. Many countries have limited the number of health specialists, either by restricting training opportunities or by restructuring physician payment to lessen the financial incentives to specialize. China might start with efforts to control the proliferation of high-technology diagnostic equipment, both through controls on the amount of equipment and through provider payment incentives.

Notes

1. The high share of spending on pharmaceuticals in China is due in part to the high price of pharmaceuticals relative to the prices of many other health inputs, such as labor. Nonetheless, there is much evidence of overprescribing.

2. See the background papers prepared for this report, especially the papers of Zhao Yuxin (1995), Cai Renhua (1995b), Chen Xiaoming (1995), Meng Jianguo (1995), Hu Haobo (1995), Hu Shanlian (1995c), and Hou Yan and Zhou Heyu (1995).

chapter six

Options for Efficient Risk Pooling in Rural Areas

In addition to strengthening funding for public health, improving the access of the poor to health services, and introducing systemwide measures to improve efficiency, China needs to improve risk-pooling mechanisms in both rural and urban areas. In China, as in other countries around the world, the distribution of health care costs is highly skewed: the 10 percent of the population with the highest health expenses accounts for about two-thirds of medical costs each year.

In China's rural areas the main insurance issue is how best to provide risk pooling for the perhaps 700 million residents who lost whatever access to prepaid health care they had under the cooperative medical system in the early 1980s. Nearly 90 percent of farm households now pay out of pocket for almost all their health services. Publicly run health institutions, now receiving minimal government subsidies, charge high user fees even for emergency services. If China allows present trends to persist, most of the rural population will have to continue to finance health services, including catastrophic care, out of pocket and pay on a fee-for-service basis. The fee-for-service payment systems will tend to escalate health spending.

Restoring and developing risk-pooling mechanisms in rural areas is now a top government priority in improving rural health services. The government can play an important part in developing health insurance by providing a policy framework that would encourage the establishment of risk-pooling mechanisms to meet the population's demand for insurance, minimize the market failures that occur in insurance, and promote forms of provider payment and health care organization and delivery that contain costs and promote efficiency over the long term.

How can China best encourage risk pooling for the general rural population? There are several possible approaches to the collective finance of a health benefit package that includes catastrophic care.

- *General revenue financing* is used in many countries, from Sri Lanka to Sweden. Under this approach, the publicly financed health system, organized either nationally (as in the United Kingdom) or subnationally (as in the Canadian provinces), finances almost all health services. In OECD countries with general revenue financing, some of the population typically purchases supplemental private insurance to cover such additional benefits as private hospital rooms. While the Chinese central and provincial governments do provide small subsidies for publicly operated health facilities, increasing these subsidies to a level sufficient to provide effective insurance coverage to the 800 million rural Chinese would be prohibitive, at least in the short to medium term.
- *Mandatory social insurance,* funded by a wage tax or premium, has also been adopted by many countries and is used in urban areas in China (the state enterprise insurance system). But social insurance has limited feasibility in rural China today because it relies on employers to enroll beneficiaries and collect contributions, and peasants are largely self-employed. In addition, China lacks the institutional and organizational capacity to manage large social insurance programs that would cover hundreds of millions of beneficiaries in rural areas.
- *Private voluntary insurance* presents several problems. Experience in developing countries has shown that a private insurance market emerges to supply health insurance only to the most affluent urban populations (Musgrove 1996). And private insurance is neither an equitable nor an efficient means for covering basic health benefits because of information asymmetry and selection bias. Risk selection by insurance companies leaves the disabled, the elderly, and the less healthy uninsured. Countries such as the United States are developing regulatory mechanisms to partially address these problems, but that regulation requires well-developed administrative and institutional capacity.
- *Community financing schemes* provide collective health financing for entire rural communities. The schemes generally derive their funds from three sources: households, government, and local industries. The local community, not the government, organizes and manages the financing and delivery of services on behalf of the consumer (Hsiao 1995b). Analysis of ongoing community financing schemes in China shows that they can lessen the inappropriate use of drugs, improve the quality of services, and reduce the overall costs of services—in part by encouraging service provision at lower-level health facilities whenever possible (China Network and Harvard School of Public Health 1996; Jin 1995a; Liu Yuanli and others 1996). Because community financing can promote universal or near-universal coverage at the local level and efficient service delivery without causing a major drain on government funds, it appears to be a more promising option for risk pooling in rural China over the medium term than the others reviewed.

This chapter looks at present and past risk-pooling schemes in rural China and examines options for reestablishing risk-pooling arrangements that will work given present economic conditions and financial and institutional capacity in China. Although China has much experience with community financing, particularly under the cooperative medical system, it is unclear how well the approach will work on a large scale today: reforms have changed the economic organization of the countryside, and the size and heterogeneity of China would complicate implementation of the approach. The approach would therefore need to be phased in and systematically monitored and evaluated. As noted in chapter 4, such community financing schemes, if they prove feasible, could also be used to channel subsidies for health care to the poorest rural residents.

The rural cooperative medical system, 1960–83

China pioneered rural, community-based health financing with the rural cooperative medical system, which operated under the agricultural commune system in the 1960s and 1970s. Under the communal system of agriculture, communes took in all farm revenues and paid them out to individuals and households on the basis of points for work provided to the commune. Barefoot doctors received points for their medical work, so in that way public health services were financed by the townships and villages with little or no subsidy from higher levels of government. Barefoot doctors delivered free preventive and primary care services at the village level. Patients typically paid a coinsurance fee for drugs, secondary services, and hospital services, with a higher fee for inpatient services.

The cooperative medical system had problems. Since it was financed largely by each village's communal welfare fund, the benefit package varied depending on the wealth of the community. Poorer communities often could afford to cover only primary care services. And often, poor administration and low levels of financing meant that schemes functioned only from autumn (harvest season, when the premium was collected) through spring of the following year. Schemes also had financial difficulties because risk pooling was only at the level of the village.

The cooperative medical system has since collapsed in 90 percent of Chinese villages. China's shift from agricultural communes to the rural production contract responsibility system weakened the collective economic foundation supporting the cooperative medical system. It eliminated the communal welfare fund, the main source of financing for the system and the only source of support for the barefoot doctors and for drugs and other health services. By 1983 only 40–45 percent of China's villages were still covered by the cooperative medical system. But most of these communities too disbanded their programs by the mid-1980s.

Even before its demise, patronage and corruption had weakened the cooperative medical system. It was controlled and managed by local officials, some of whom used their power for selfish gains. As a result, farmers lost confidence in the government-run cooperative medical system and refused to pay premiums once the system became voluntary after the late 1970s. This experience underscores the importance of effective organization and management of any new community financing schemes and the need for an adequate financial base.

The government did not replace the cooperative medical system with a new health financing structure, but instead adopted a laissez faire policy. Many communities designed new funding mechanisms. Many others fell back on a system of fee-for-service payments. And some allowed farmers who could not afford to pay for health services to obtain support from the village welfare fund. Such funds were financed by a specific tax of just under 5 percent of the village's net farm production. A fifth of this agricultural tax was designated for welfare assistance and to defray the health care costs of those in need.

China's recent experiences with community financing

China can also look to its more recent experiences with small-scale community financing schemes to identify sound policy options. These include the completed Sichuan Rural Health Insurance Experiment (see Mao 1995; Sine 1994; Cretin and others 1995) and the ongoing World Health Organization (WHO) Fourteen Counties Study of Community Financing (Yang 1995). In addition, the Study of Thirty Poor Counties produced a wealth of data on rural health finance frequently cited in this report (Luo 1995; China Network and Harvard School of Public Health 1996). Surveys done in preparation for the World Bank–supported Rural Health Workers Development Project also produced important information on community financing schemes in China today.

Sichuan Rural Health Insurance Experiment

The Sichuan Rural Health Insurance Experiment, which took place in 1989–90, assessed the potential effect of insurance and coinsurance on the demand for health care and estimated the likely costs of providing services under an insurance regime. This experiment involved a sample of twenty-six villages in two counties and 40,443 people (Sine 1994).

Three insurance benefit plans were implemented with varying reimbursement rates for inpatient and outpatient services. Premiums were set at 1.5 percent of average income. Insured individuals could freely visit village and township facilities but could visit county hospitals only in an emergency or with the approval of the township health center. The experiment showed that households were willing to join such a scheme—more than 90 percent of households in the test areas voluntarily joined the program and 95 percent reenrolled after the first year. It also showed that administrative costs could be kept low (8 percent of total reimbursements). The study also found that:

- Coinsurance (the requirement that the patient pay part of the cost of health services at the point of service) exerted a significant negative effect on demand for care across different population groups. There were no interactions between the effect of coinsurance and age, income, or health status.
- Users in all but one village reported high satisfaction with the insurance arrangement (Mao 1995, p. 16).
- Services were used less when there was no functioning village health station, underlining the importance of an adequate supply of basic services (Mao 1995; Sine 1994).
- As in other countries, a small share of the population accounts for a large share of total health expenditures, underscoring the need for catastrophic insurance. About 11.5 percent of the covered population incurred 70 percent of the total health expenditures.

WHO Study of Fourteen Counties

To learn how to improve organization, financing, and service delivery, the World Health Organization study is assessing community financing schemes in fourteen counties, in Beijing, Henan, Jiangsu, Zhejiang, Jiangxi, Hubei, and Ningxia. In each county a research team has interviewed 540 households and surveyed health services.

The study has found that a typical community fund might collect 5 yuan per person from families, 1 yuan per person from the village's social welfare fund, and 1 yuan per person from the township. Patients typically must pay a deductible (for example, 100 yuan) and make a copayment on expenditures above the deductible. The schemes limit coverage of drugs to 120 kinds of medicine, including traditional Chinese medicines, and set a limit on reimbursement for diagnostic tests.

Study of Thirty Poor Counties

The Study of Thirty Poor Counties was conducted in 1993–95 by a network of Chinese universities and Harvard University. It found that 16.5 percent of the villages it surveyed still maintained some type of community-based health finance scheme, covering 11.6 percent of the sampled population (table 6.1). About two-thirds of the schemes covered only primary care services at the village level; a third covered comprehensive services, ranging from primary care to inpatient services. The benefit structures all incorporated coinsurance and often set high copayment rates for inpatient services. The study found that the most prevalent type of community fund management was by village committee or by the village and township jointly (table 6.2).

The study appears to provide strong support for reestablishing community-financed health care. As part of the study, 11,044 randomly selected households were asked about their preferences for such organized financ-

TABLE 6.1
Prevalence and benefits of community health financing in thirty poor counties, 1993

Type of benefits	Number of villages covered	Percentage of villages covered	Percentage of population covered
Comprehensive	29	5.1	4.4
Primary care services only	59	11.4	7.2
Total	88	16.5	11.6

Note: Comprehensive benefits refers to schemes that reimburse 30–100 percent of hospitalization fees for township and county-level hospitals and 50–100 percent of outpatient fees. *Primary care services* refers to coverage of fees (or discounted prices) for most village-level services with fees at the township and county levels paid out of pocket by patients.
Source: China Network and Harvard School of Public Health 1996.

TABLE 6.2
Management of community health financing in thirty poor counties, 1993
(percent)

	Type of benefits	
Form of management	Comprehensive	Primary care services only
Township government	17.2	3.4
Township health center	20.7	6.8
Village and township jointly	20.7	10.2
Village committee	34.5	47.5
Village and township doctors	6.9	32.1

Source: China Network and Harvard School of Public Health 1996.

ing schemes. Seventy percent of the households not covered by community financing or child immunization and maternal and child health care prepayment schemes responded that they would like to see an improved scheme similar to the cooperative medical system established; 88 percent of those covered by community financing schemes stated that they would like them to continue. Among the households without coverage that favored reestablishing community financing schemes, about a fourth preferred to see such a scheme managed by the village, a fourth preferred that it be managed by the township, and the rest preferred joint management by the township and village or management by the health facility.

Another indication of management preferences was obtained in the course of preparing the World Bank–financed Rural Health Workers Development Project. More than 1,000 village cadres in five provinces were asked about their preference for the management of community-financed health schemes. Close to 60 percent preferred management by a village committee, and roughly a fourth preferred joint management by the township and village.

In most rural areas, particularly poor areas, it is not possible to derive adequate revenues for any organized financing scheme solely from households. Funding must come from multiple sources. The Study of Thirty Poor Counties and the Five Province Survey done in preparation for the Rural Health Workers Development Project found that about half the financing for existing community-financed health plans came from household contributions, about a fourth from village social welfare funds, and about 10 percent from the government (table 6.3).

TABLE 6.3

Community health financing by source in selected counties and provinces, 1991 and 1993

(percent)

	Government	Village social welfare fund	Households	Other
Funds surveyed in Study of Thirty Poor Counties (1993)	16.1	20.3	48.1	15.5
Funds surveyed in Five Province Survey (1991)	8.0	30.3	58.7	3.0

Source: China Network and Harvard School of Public Health 1996; World Bank 1993a.

The Five Province Survey—which covered Hebei, Shanxi, Fujian, Guizhou, and Henan—found that the relatively poor province of Shanxi had the greatest coverage—close to two-thirds of the villages maintained some form of community financing (table 6.4). But in another poor province, Guizhou, very few villages had community financing (0.8 percent). In these poor provinces the schemes were financed largely by household contributions; and benefits covered only primary care services because of the small contributions that poor households were able to make.

A new policy direction for financing rural health care

On 2 July 1994 the Chinese government announced a new policy direction for the financing of rural health care in a front-page article in the *People's Daily.* The new policy appeared to draw in part on lessons from China's experience with the cooperative medical system

TABLE 6.4

Prevalence and benefits of community health financing in five provinces, 1991

(percent)

Province	Number of villages with community financing	Share of villages with community financing	Share of schemes with comprehensive coverage	Share of schemes with primary care services only
Hebei	3,992	13.1	42	58
Shanxi	4,727	65.6	15	85
Fujian	512	6.3	25	75
Guizhou	160	0.8	6	94
Henan	1,590	6.2	7	93
Total	10,981	12.2	24	76

Note: Data from the Study of Thirty Poor Counties indicate that almost 80 percent of the category "services and drugs coverage" is comprehensive coverage. Therefore, for the purposes of this table, the Five Province Survey data were recategorized, with services and drugs plans counted as comprehensive coverage and the remaining categories (services only, drugs only, other) counted as primary care services coverage only.

Source: World Bank 1993a.

and other community financing schemes. The government called for the development of community-based schemes to fund and organize health care for the rural population, guided by the following principles:

- The government's role is to establish policy and provide leadership.
- Each community organizes its own collective financing for basic health care.
- Funding will be derived from multiple sources (government, collectives, and individuals).
- Priority should be given to covering preventive services.
- The schemes and benefit package should vary according to community conditions and economic capacity.
- The schemes should be supervised by and accountable to the people.

The 1994 policy announcement emphasized that participation in community financing should be voluntary rather than compulsory for individuals. The government also suggested three management models with varying degrees of community control.

Although these health funding principles are supported by the State Council, there is wide variation among provinces in the interpretation of the policy. There is confusion, for example, about the Ministry of Agriculture's decree that the tax burden on farmers must be reduced. Some provinces view this decree as constraining any community-financed schemes that involve household contributions. As a result, the government's support of community-financed health care has had little impact.

China's villages and townships would have stronger incentives to develop community-based collective financing if the central government clarified and elaborated its priorities. The government could also provide technical assistance and perhaps a subsidy for establishing and operating community financing. There are strong economic arguments for such a subsidy as a means to encourage equitable and efficient risk-pooling mechanisms in rural areas.

What would it take to make community financing work?

Several basic elements are needed to establish equitable and efficient community financing.

An affordable and acceptable benefit package

The first major issue in developing a community financing scheme is to determine what benefit package and coinsurance levels are acceptable and affordable for most community residents. While economic principles such as cost-effectiveness should help guide the design of the benefit package, the end product must have popular support and people must be willing to pay their share of the required premium. The Chinese Network of Health Economic Institutions developed several potential benefit packages to test consumer demand and costs. The prototype test packages cover the full range of services, but with relatively high coinsurance rates for drugs, demand-elastic services, and inpatient hospital care. In aggregate, the proposed coinsurance rate is about 30 percent. The estimated cost for these illustrative benefit packages is 30 yuan per person a year in 1993 prices, equal to 5 percent of the disposable income of rural households with modest incomes (box 6.1).

The Sichuan Rural Health Insurance Experiment and household survey responses indicate that rural residents want coverage for a wide range of products and services, from drugs to village doctors to county hospitals. They also seem to accept coinsurance payments. Field tests could reveal whether households are willing to make voluntary payments to cover the expense of providing the desired services.

Choice of providers

To promote consumer satisfaction and competition on the supply side, beneficiaries should be allowed to choose their primary care provider wherever possible. If beneficiaries seek services from qualified providers not included in the scheme, the collective financing scheme could reimburse them at a reduced rate.

Universal coverage and adequate risk pools

If possible, enrollment should be mandated at the local level to minimize adverse selection. In designing schemes, consideration should be given to an adequate population for risk pooling at the village and township level. The population of a village generally averages 1,000 people—a relatively small size for pooling hospi-

BOX 6.1

Estimating the costs and content of a benefit package for rural China

Hsiao and others (1996) used household survey data, financial records of health facilities, and the results of the Sichuan Rural Health Insurance Experiment to model the likely utilization and costs of a basic package of services under a community financing scheme in rural China. Since most of the rural population is uninsured, the provision of a basic benefit package would increase the demand for services. To estimate the demand increase, the authors used demand elasticity estimates from the Sichuan Rural Health Insurance Experiment, which assumed that a 10 percent increase in insurance coverage would increase outpatient expenditures by 6 percent and inpatient expenditures by 4 percent. They assumed that supply is perfectly elastic because China has an excess supply of personnel and beds at the village and township level. They also assumed that the community financing scheme would pay for inpatient services on a case basis, with payment levels determined prospectively, rather than on a fee-for-service basis, and that this payment system would generate 15 percent savings in inpatient costs.

The basic benefit package modeled by the authors covers services at the level of the village, township health center, and county hospital. The package includes catastrophic care, but such care would be limited by what the county hospital can provide. (A county hospital is not equipped and staffed to carry out open-heart surgery, for example.) Two prototype packages were developed, with different levels of coinsurance. The medium level requires a 30 percent copayment for township health center outpatient fees, a 40 percent copayment for drugs, and a 35 percent copayment for inpatient charges at the county hospitals. The high level requires a 30 percent copayment for township health center outpatient fees, 50 percent for drugs, and 45 percent for inpatient charges at the county hospitals. A stop-loss level for coinsurance was built in at 500 yuan. The package with a high level of coinsurance was estimated to cost 28 yuan per person, and the package with a medium level 31 yuan per person.

Some economists argue that health insurance should cover only catastrophic expenses. But while risk pooling that excludes catastrophic expenses does not make sense, there are sound arguments for covering both catastrophic and some noncatastrophic expenses in a risk-pooling arrangement. Insurance that covers only catastrophic care creates incentives to provide care in a hospital setting when less costly but often equally effective outpatient care might be available. And it creates disincentives for consumers to seek early treatment, which is typically more cost-effective.

tal expenses. A township generally averages 12,000 people, a population adequate for pooling the risk of hospital expenses. The benefits of a larger risk pool need to be weighed against the decline in accountability that occurs as the size of the pool increases.

Effective referral and supervision

A financing scheme would need to establish effective referral systems between village doctors, township health centers, and county hospitals. And it would need to include provisions for monitoring the quality of care.

Provider payment systems with the right incentives

The method of paying health workers affects the overall cost and quality of care. Strong consideration should be given to replacing fee-for-service payments with a salary plus performance bonus for village doctors. This payment method would minimize incentives for village doctors to overprescribe tests and drugs. It would also reduce administrative expenses by eliminating the need for processing claims. For hospitals, capitated payment contracts for services would strengthen incentives to provide services efficiently to the covered population.

Transparent and accountable management

Misuse of funds, favoritism, nepotism, and, in the worst cases, corruption will destroy the solidarity necessary to make community financing work. Ideally, community-based financing schemes would be nongovernmental entities with directors elected by those enrolled in the schemes, and they would be required to provide frequent financial and quality-of-service reports. Such a system would give residents of a community a strong sense of control over their community's financing scheme.

Adequate financing

The sources of financing for such schemes could include, in varying proportions, households, rural collective enterprises, village social welfare funds, and central, provincial, and local government subsidies. In the Sichuan Rural Health Insurance Experiment farmers paid 1.5 percent of their incomes in premiums to buy health insurance, in addition to making copayments. Rural collective enterprises voluntarily fund community health in some richer villages. Village social welfare funds devote about a fifth of their spending to health. Local government subsidies would vary with local conditions and with the priority given to health care by the local government.

Government support

To provide an incentive for communities to establish collective financing schemes, the central government should consider providing technical assistance to help them do so rather than resorting to politically unpopular mandates. It could also consider providing a matching grant to supplement the contributions of households, rural collective enterprises, village social welfare funds, and local governments. The subsidy might go only to townships in which collective financing schemes meet certain basic government guidelines. There is an efficiency argument for such a government subsidy, because the incentive would promote the development of efficient risk-pooling mechanisms.

The central government financing might initially be 5 yuan per person for a 30 yuan-per-person benefit package for nonpoor townships (and considerably more for poor villages, as discussed in chapter 4). This subsidy would grow with income. By 2001, if 120 million people in rural areas received the matching subsidy, the cost to the government might be 1.2 billion yuan (with a 10 yuan per capita subsidy).

China's experience with community financing indicates that it may be a promising approach for reestablishing risk-pooling arrangements for catastrophic medical expenses in the country's rural areas. Depending on the design of the scheme, community financing can also promote efficient service delivery. Nonetheless, China is a large and heterogeneous country, and while community financing has many advantages in theory, it may prove difficult to administer on a widespread basis. It may be difficult for a community financing scheme to collect premiums, administer contracts, and remain solvent, for example, especially in a poor community with little administrative capacity. The approach therefore needs to be phased in with technical assistance and systematically monitored and evaluated. While some general guidelines are important, local experimentation and adaptation should also be encouraged. Alternatives could be tested for several key design elements:

- Benefit packages, including coinsurance levels, to test consumer demand and costs.
- Provider payment arrangements (such as a salary plus performance bonus for primary care providers and capitation payments for hospital stays and visits).
- Management models.
- Simple methods for monitoring the quality and appropriateness of care, including the prescribing of drugs.

chapter seven

Options for Efficient Risk Pooling in Urban Areas

In sharp contrast to the very low health insurance coverage in China's rural areas, coverage extends to about half the population in urban areas. Nevertheless, there are pressing problems in the urban insurance systems. The two formal insurance systems, the government and state enterprise systems, cover just 15 percent of China's population yet account for 36 percent of health spending. There are many problems in their operation, particularly in their financial sustainability, their impact on the market, and their efficiency. The systems are in urgent need of reform to avoid becoming an obstacle to economic modernization and state enterprise reform. But change will be difficult because of the many powerful groups with vested interests in these insurance plans. Gradually expanding coverage to uninsured groups in urban areas will be another challenge.

Beneficiaries of the government system, particularly retired veterans but also civil servants and university students, oppose any reduction in their benefits. Enterprises and government units with relatively young workforces or few retirees oppose pooling their risks with employers of primarily older workers. Government hospitals oppose payment reforms that may reduce their revenues or force their closure or downsizing. Ministries also defend interests, which are often conflicting. The Ministry of Finance is concerned mainly with containing government outlays. The State Pharmaceutical Administration wants to increase drug sales. The Ministry of Health wants to protect its hospitals' revenues. The Ministry of Labor defends the interests of China's state enterprises, which themselves operate thousands of health facilities. Closing hospitals and firing staff is just as unwelcome as closing a steel plant or a coal mine. It is not a viable policy option except at the margin.

These constraints require a balancing act. The government needs to promote policies that control beneficiary demand for health care, improve efficiency and reduce waste in hospitals to keep costs down, yet allow hospitals to earn enough to cover their costs

Experiments in reforming the urban health insurance systems

Several urban health finance experiments—in Shenzhen, Shanghai, and Zhenjiang and Jiujiang—are identifying strategies for national reforms of the government and state enterprise insurance systems and, eventually, for broadening urban insurance coverage. These experiments focus on improving incentives in the system, particularly through provider payment reforms and coinsurance. The most recent experiments, in Zhenjiang and Jiujiang, appear to hold the greatest promise as national models. As a result, in January 1996 the government decided to replicate the Zhenjiang and Jiujiang reforms in two additional prefectures or cities in each province, for a total of roughly fifty sites.

Mixed results from the Shenzhen experiment

Shenzhen is a newly established city of 3 million people—1 million permanent residents and 2 million temporary residents who tend to be young contract workers. In 1995 Shenzhen implemented a new social health insurance program with the dual goals of extending insurance coverage and reducing health costs. The scheme is managed by the Shenzhen Bureau of Health Insurance, which is controlled by the Social Insurance Bureau.

All employers were required to enroll their workers in the program. The insurance program provides a two-tier benefit structure. The first tier is an individual medical savings account in which the employer deposits 6–10 percent of a worker's average annual wage (varying by age). This account can be drawn on only to pay for medical expenses. If expenditures exceed the amount in the individual savings account, the patient is reimbursed from the second tier of financing—the social risk-pooling fund. For these expenses the worker must pay a 10 percent coinsurance fee, up to a stop-loss ceiling set at 8 percent of the wages that the worker earned in the previous year. Beyond this ceiling the social risk-pooling fund pays 100 percent of the worker's medical bill. The balance remaining in an individual account at the end of the year is carried over to the next year, as is the balance in the social risk-pooling fund.

Shenzhen finances its social health insurance through a payroll tax on employers set at 8–10 percent, depending on the age and sex composition of the workforce. Workers do not contribute directly. Benefits for retirees are financed from pension funds, which pay 10 percent of the monthly retirement benefits to the health insurance fund as a premium.

The payment system can be characterized as fee-for-service with a cap. Providers bill the insurance fund for each item of service. Caps are set separately for inpatient and outpatient visits. On a quarterly basis the insurance fund reviews the bills submitted by providers and calculates payments that include a bonus for bills below the cap and a penalty for bills exceeding the cap. To ensure quality, 5 percent is withheld until the provider passes certain quality performance standards.

The Shenzhen program has had difficulties meeting its stated goals of expanding coverage and reducing costs. Enrollment and premium collection have been problematic. Joint ventures and privately owned companies have refused to participate in the social insurance program, and the Bureau of Health Insurance has found it almost impossible to enforce their participation, which is compulsory under the regulation establishing the program but not under any law. Consequently, only 5 percent (150,000) of the targeted population of 3 million have participated. Almost a third of the enrollees (46,000) had been covered by the government insurance system, and most of the rest by the state enterprise insurance system. Only 18 percent of enrollees (27,000) had been uninsured. The new program also enrolled some 8,900 retirees.

Shenzhen has also encountered financial difficulties, both because of adverse selection in enrollment and because of fraud. Employers who have refused to enroll in the program disproportionately employ younger workers, as shown by the fact that the average age of enrollees (thirty-six) exceeds the citywide average (twenty-seven) by nine years. There is also evidence that many employers lower their premium contributions by underreporting wages. The underreporting is conservatively estimated at 50 percent.

Shenzhen has had problems, too, in controlling hospital costs. The Bureau of Health initially controlled the Shenzhen Bureau of Health Insurance, which meant that it both managed the health facilities and controlled their financing. As a result, the conflict of interest between providers and the insurance administration was resolved mostly in favor of the providers, at the expense of cost control.

Shenzhen's social insurance program could not maintain solvency with the Bureau of Health Insurance acting as both funder and provider. Because of this conflict of interest, the Shenzhen Bureau of Health Insurance is now controlled by the Social Insurance Bureau.

Global budgets to control health costs in Shanghai

Shanghai is a metropolis of about 14 million people. Seven million of the city's workers are, in principle, covered under the government and state enterprise insurance plans, and perhaps another 3 million people are covered as dependents under these plans. Shanghai faces problems in its health sector similar to those of other urban areas—high cost inflation, poor quality of services, a large share of uninsured residents, and little risk pooling.

In the early 1990s Shanghai developed a plan, yet to be implemented, to overhaul its health care financing system. Under this plan the government and state enterprise systems would be combined into a single citywide social health insurance plan. Copayments would increase: insured workers would pay 15 percent of outpatient charges (up to a maximum of 1,000 yuan) and 8 percent of inpatient charges (up to 2,000 yuan). The plan includes a medical savings account for individual workers (4 percent of wages) and a social health insurance fund (15 percent of wages), with most of the contributions to come from the employer. The social insurance program would be managed by the Shanghai Bureau of Health, which would establish a health insurance bureau to handle both the insurance program and the pension scheme. Employers would manage the individual accounts, from which workers could withdraw funds equivalent to the coinsurance they pay for health services.

The long-term goal of the plan would be to expand coverage to workers employed by joint enterprises, smaller collective industries, and private enterprises. Workers in the "big eight" industries (including railroads, airlines, coal, steel, and the postal service) would be permitted to set up their own systems.

Shanghai gave first priority, however, to adopting a new payment system to control hospital cost inflation. From 1 July 1993 through 30 June 1994 hospital costs had risen 53 percent. In July 1994 Shanghai implemented a global hospital budget that would permit total costs to rise no more than 24 percent over the next year, and drug expenditures no more than 15 percent. It also reformed prices. Fees for visits and surgeries were increased to more accurately reflect the labor costs for these services, and fees for CT and MRI services were reduced by 12–15 percent. The hospital continued to be paid on a fee-for-service basis. At the end of September 1995 Shanghai found that its global budget had kept hospital cost increases within the target rates.

Promising preliminary results from the Jiujiang and Zhenjiang experiments

A demonstration health insurance project sponsored by the State Council began in December 1994 in Jiujiang and Zhenjiang, two medium-size cities on the Yangzi River in Jiangxi and Jiangsu Provinces. These cities, each with about 2.5 million inhabitants, form the "cutting edge" in urban health finance reform, and they have been visited by more than 600 officials from other provinces and municipalities.

In both cities newly organized insurance centers collect insurance payments from enterprises and public agencies, then commit these funds to individual and group accounts by formula (figure 7.1). The centers draw on these accounts to pay medical fees. The personal medical savings accounts plus large copayments should help moderate patient demand while providing stop-loss coverage to protect against catastrophic medical bills (box 7.1). The insurance centers also introduced an essential drug list of 1,100 Western and 500 traditional Chinese medicines for which they will provide reimbursement.

The Zhenjiang model includes two additional innovations. First, it has combined the government and state enterprise insurance systems into a single insurance center. Second, it sets packaged, or bundled, fees per outpatient visit and per inpatient admission, establishing the payment rates prospectively.

All enterprises in Jiujiang and Zhenjiang that had government or state enterprise insurance were asked to join the pilot study. By the end of August 1995, 95 percent of the eligible population in Zhenjiang had joined and 93 percent of the contributions had been collected. In Jiujiang 90 percent of the government units and 88 percent of the enterprises had paid their contributions. In total, 415,000 people in Zhenjiang and 370,000 people in Jiujiang are enrolled in the new plans.

FIGURE 7.1

Wage tax flows to individual and common accounts in the Jiujiang and Zhenjiang experiments

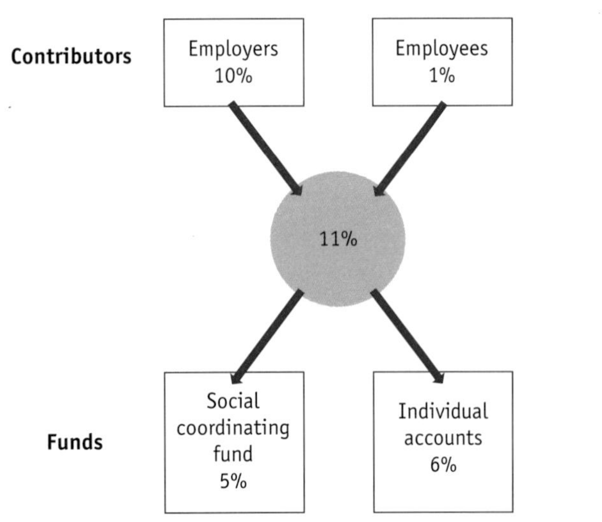

Although a thorough evaluation is still needed, results from the first year of operation appear positive, with evidence of successful cost containment and reasonable control of administrative costs.

- Coverage gaps, which had been especially large among teachers and workers in deficit-ridden enterprises, were eliminated.
- The rates of overprescription and use of expensive diagnostic tests were cut significantly.
- The annual rate of growth in aggregate hospital expenditures declined by 23–28 percentage points from the average annual rate of 33 percent in 1991–94.
- The quantity of services declined 9 percent, and the bed occupancy rate 2 percent, compared with the previous year.
- Both inpatient and outpatient visits fell for enrollees (Cai 1995b; Zhenjiang 1995; Yip 1996).

BOX 7.1

The three tiers of financing for health services in the Jiujiang and Zhenjiang experiments

The Jiujiang and Zhenjiang experiments both finance health expenditures through three tiers of financing. The first tier is the individual account. About 6 percent of a worker's annual salary is deposited in this account, with some variation in the rate by age, so the size of the account depends mainly on the worker's salary.

The second tier consists of out-of-pocket expenditures for medical bills that exceed the amount in a worker's individual account. When a worker's health expenditures exceed that amount, the worker pays a deductible of up to 5 percent of his or her annual income before receiving reimbursement from the third tier of financing.

The third tier is the social risk-pooling fund, designed to insure workers against the financial burden of catastrophic illness. The larger the expense (beyond the individual deductible), the more the social risk-pooling fund pays, with slightly more generous coverage in Zhenjiang.

Consider a worker with an average annual wage of 6,000 yuan. If the worker incurs medical expenses in a given year that total less than 6 percent of his or her annual salary (that is, less than 360 yuan), the expenses are fully paid from the individual account. If the worker has medical expenses of 560 yuan in that year, the individual account pays 360 yuan and the worker pays the remaining 200 yuan from current income as a deductible. But if the worker's medical expenses total 15,000 yuan, the social risk-pooling fund will pay almost 90 percent of the bill. For this catastrophic medical expense—equal to two and a half years' income—the fund limits the worker's financial loss to about a quarter of his or her annual salary.

In early January 1996 the State Council decided to replicate the Jiujiang and Zhenjiang experiments in other cities throughout China, even before formal evaluation.

The Jiujiang and Zhenjiang experiments suggest that there are clear benefits from separating the funding of health services from their provision. This separation can encourage cost containment, efficiency, and service quality because the funding agency represents the interests of the consumers, not the health providers, and can bargain with hospitals and monitor service quality. So far, the Zhenjiang insurance center has focused on innovative payment methods for hospital services (packaged fees). In the future it also needs to monitor the quality and appropriateness of care. The Jiujiang insurance center continues to pay hospitals on a fee-for-service basis and tries to monitor the appropriateness of care by reviewing claims.

Expanding the Jiujiang and Zhenjiang experiments

Further evaluation of the ongoing Jiujiang and Zhenjiang experiments is needed, and as they are replicated in all provinces, all the experiments need to be systematically monitored and evaluated to maximize learning. The evaluation should focus on several questions: How do the individual accounts and deductibles affect patients' demand for services? How do payment caps and global hospital budgets affect the quality of medical services? How do hospitals alter their behavior under the reform, including in efficiency, cost shifting, risk selection, and capital investments? What is the appropriate share of wages to allocate to the individual account? And what organizational changes are induced by reforms, such as in referrals between health facilities?

As the demonstration projects are extended to fifty more cities and prefectures, program managers could consider these potential improvements:

- Expand coverage to include dependents.
- Protect the funds accumulated in the individual accounts and the reserves for the social risk pool by paying inflation-adjusted rates of return.
- Establish modern scientific accounting and auditing methods, including public quarterly financial reports for the insurance centers. The operations of the insurance centers need to be as transparent as possible to win the confidence of beneficiaries, employers, and providers.
- Include representatives of employers, workers, and providers in each city on a board of trustees to oversee the operations of the insurance program.
- Direct insurance centers to pay hospitals on a case and packaged-fee basis, using the categories of, say, trauma, surgery, medical, cancer treatment, pediatrics, and long-term infectious diseases.
- Adjust prices for labor inputs, drugs, and diagnostic tests to end the distortions now causing much of the obvious waste in the health system.

One aspect of the insurance experiments deserves rethinking. Under the government and state enterprise insurance systems all wage tax contributions were available for risk pooling, and a person who used no health services received no benefits while a worker with major medical costs paid nothing. But in the Jiujiang and Zhenjiang experiments benefit costs and individual health accounts transfer income from the sick to the healthy worker. Under these experiments about half the wage contributions are deposited in individual accounts—much of which will never be spent on health care because many people never use health services. Meanwhile, very ill workers will exhaust their individual accounts, paying another deductible equal to 5 percent of their wages before the risk pool begins paying a portion of their medical expenses. In its evaluation of the experiments the government should weigh the acceptability of such outcomes against the cost savings generated.

Links between pension and health insurance reforms and economic modernization

Labor mobility has until recently been extremely limited in China. State enterprise workers, once hired, stayed with their employer for life. In turn, the state enterprises provided their workers with pension and health benefits on a pay-as-you-go basis. With lifetime employment, there was little need to make pensions and health care rights portable, although unfunded liabilities for pensions and retiree health benefits are already undermining the viability of pay-as-you-go finance. Further, linking pension and health benefits to employment makes workers dependent on their employers, not only for their jobs but in order to realize their pension rights and to have health insurance both while working and after retiring. Survival of the enterprise then becomes a political necessity, even at the cost of open-ended public subsidy.

To improve efficiency in the state enterprise sector, noncompetitive enterprises must be allowed to fail or restructure, and labor must therefore become more mobile. Delinking pension and health benefits from the enterprise helps to make this possible. The present system is an obstacle to economic modernization, which requires labor mobility and state enterprise reform. Pension and health finance reforms are needed to ensure that workers can change jobs without jeopardizing their pension and health benefits, to deal with problems of pay-as-you-go finance and the cost of health care for retired workers, and to facilitate enterprise reform.

China appears to be moving away from a pay-as-you-go pension system operated by individual enterprises to a unified pension system in which a substantial part of retirement income comes from fully funded individual accounts. Benefits need to be reduced to make the new system financially sound. Coverage could be extended by gradually including all formal sector workers in urban areas and employees in large township enterprises.

This pension reform would produce important benefits: it would delink pension administration from enterprise management, encourage labor mobility, contribute to capital accumulation, provide incentives for saving, provide protection against poverty, and contribute to income security in old age (for covered workers). One issue in implementing such reforms is the cost of financing the transition, given the large outstanding unfunded liability of current pensioners and the accrued pension rights of current workers. In addition, without parallel reform of health finance, it is unlikely that pension reform can meet the objectives of facilitating labor mobility and state enterprise reform. Reform of enterprise health finance should therefore proceed simultaneously with (and on similar principles as) pension reform.

Options for broadening urban risk pooling

In addition to the ongoing experiments with the government and state enterprise insurance systems, health insurance coverage needs to be broadened to include the uninsured in urban areas. Providing full coverage for dependents under the government and state enterprise systems would significantly expand insurance coverage in urban areas. But over the medium to long term some form of risk pooling needs to be developed to cover the growing share of the labor force that works outside the state-owned sector in urban China, as well as their dependents and the elderly. As the Shenzhen experience shows, getting joint ventures and privately owned companies to participate in social health insurance will be difficult. It will take many years to set up well-functioning insurance systems that cover the majority of the urban population. In the meantime government support to public health and to basic services for the poor will help to meet the needs of the urban residents not covered by insurance.

Various combinations of public and private financing arrangements could be developed for risk pooling in urban areas. For the workers outside the state-owned sector, the government could encourage individual participation in an expanded state enterprise insurance system, complemented by public support for the indigent. The resulting arrangement would be akin to the social insurance systems for health care that now serve much of Europe. To fully insure the dependents of those now insured, as well as the employees of private enterprises and joint ventures and their dependents, insurers (public or private) will need to be established. These insurers could be modeled on the Jiujiang and Zhenjiang insurance centers.

A number of intermediate steps could be taken to support the long-term goal of establishing broad-based, equitable, and efficient urban risk pooling. Some of these intermediate steps are already being tested in the Jiujiang and Zhenjiang insurance experiments.

One option is to take a laissez faire approach over the short to medium term and let private, voluntary insurance emerge, with some government regulation, to close part of the gap in urban coverage. Employers could choose to offer health insurance and either self-insure or join a larger insurance pool. Individuals could join their employer's plan, if there is one, or voluntarily purchase private health insurance. There are problems with this model, however. Small employers would have difficulty purchasing private insurance. Workers would risk losing their insurance if they changed jobs. And people with high health risks would have difficulty obtaining coverage.

Another option is to require firms to offer health insurance to their workers, without requiring the firms to pay for it. That would at least provide the advantages of some risk pooling. Alternatively, the government could mandate employer and employee contributions for health insurance. Initially, health insurance could be purchased by the firm. Ultimately, mandated contributions might be pooled in one fund, with the government contributing money to cover the indigent and elderly.

These options for expanding risk pooling need to be carefully reviewed, especially in the light of the experience in Jiujiang and Zhenjiang. Intermediate steps will be needed to phase in coverage and build up administrative and regulatory capacity. Implementing reforms will inevitably require some public financial support for studies, technical assistance, and monitoring and evaluation.

chapter eight

Recommendations and Implications for Public Finance

China had an enviable record in improving the health of its people between 1950 and 1980. But the move toward a market economy that began in 1978 fundamentally altered its financing and organization of health care. Without a coherent national health policy that appropriately defined the roles of the government and the market in the reformed Chinese economy, the health system drifted, reacting ad hoc to the nation's changing economic and social environment. Now China faces several major health care problems (table 8.1).

How the Chinese government addresses these health issues will strongly affect health conditions, equity, risk sharing, efficiency, and total health care costs. International experience suggests that China's health policy is at a critical juncture—and that if China continues on the present course, the problems will deepen and become more difficult to remedy. When the United States, for example, failed to address its risk-sharing problems in the 1940s, it left the people with

TABLE 8.1
Major health care issues in China

Issue	Likely causes
Health status concerns—Such health status indicators as life expectancy are relatively high and continuing to improve. But health status remains poor among some population groups, and some data suggest that child mortality rates may have stagnated in recent years despite rapid improvement in standards of living.	• Government spending for public health programs has declined. • Income inequality has grown. Some 100 million rural residents remain very poor. • Access to health care has declined as a result of inadequate public financing for the poor and lack of organized financing for the rural population.
Inequitable access to health care.	• Rapid health cost inflation makes services less affordable to many. • As a result of the lack of a rural health financing scheme since the collapse of the cooperative medical system, many poor and near poor forgo necessary medical services. • Much of the urban population lacks insurance coverage, including unregistered migrants, workers in private or collective enterprises and joint ventures, and the workers' dependents.
Inefficiency and waste—These conditions have led to many problems, including a decline in clinical effectiveness and in the quality of services.	• The allocation of public resources favors hospital services over public health spending. • Distorted prices encourage overuse of drugs and high-technology diagnostic tests. • Overlapping hospital systems and vertical health programs lead to duplication of facilities. • Clinics, hospitals, and epidemic prevention programs all rely on user fees to finance their operations, which encourages overprovision.
Rising health care costs—Real per capita health spending increased an average 11 percent a year in 1986–93.	• Distorted prices encourage overprovision of drugs and expensive diagnostic tests. • Fee-for-service payment encourages overprovision. • The aging of the population and increasing incidence of chronic diseases raise costs.

the worst health risks—the elderly, the disabled, and the poor—uninsured by the private market. Eventually, the government had to bear the burden of insuring these groups, yet 15 percent of the population remains uninsured today. Meanwhile, health costs have escalated, consuming some 14 percent of GNP in 1995. The Republic of Korea and the Philippines failed to correct their fee-for-service payment systems and now too face rapid inflation of health costs. By contrast, Canada and Germany took timely and effective action when problems appeared in the 1970s and 1980s. As a result, they were able to ensure equitable access to health care for their citizens, relatively low health cost inflation, and steadily improving health conditions. Many other OECD countries have also undertaken health reforms in recent years, which may reveal useful lessons for China (box 8.1).

Priorities for government health spending

As a general principle, governments should not spend their limited resources on health services that people will readily buy for themselves. Tax revenues are difficult to collect, and government spending is often less effective in meeting individual demand than spending by individuals themselves. These considerations underline the importance of using public sector funds for important goods and services that would not—or cannot—be purchased privately, such as cost-effective public goods and priority clinical services for the poor. A related principle is that government health spending and regulation should seek to induce effective and efficient use of health spending by other providers and buyers of health services. In China direct government spending on health, including that through the government insurance system, was no more than a quarter of total sector spending in 1993. Both individual purchasers of health services as a group and state enterprises as a group spent far more on health than did the Chinese central, provincial, and local governments.

Following these principles and to address some of the health care issues in table 8.1, the Chinese government will need to substantially increase funding for public health programs—the activities that the government

BOX 8.1

Lessons from health system reform in OECD countries and Singapore

OECD countries have used a wide variety of arrangements for financing and delivering health services and paying health care providers. Many of these countries have undertaken incremental reforms in their health systems over the past fifteen years. These reforms have been aimed at achieving universal or near-universal access to care, containing costs (and improving efficiency), and, more recently, improving consumer satisfaction. What lessons can China learn from these reforms?

Extending coverage and pooling risk

- Except for Mexico, Turkey, and the United States, all OECD countries have achieved universal access to health care, largely financed through national and local governments or mandated social insurance.
- It is difficult to achieve universal coverage in systems that rely heavily on private voluntary insurance, because private insurers will compete on the basis of risk selection and people with higher-than-average expected health expenses will be unable to obtain insurance. Private voluntary insurance can also impede labor mobility if workers are reluctant to change jobs because of a risk that they would lose their health insurance coverage. Some of these problems can be alleviated through regulation.
- An alternative to mandating health insurance is to mandate savings that can only be used to pay medical expenses—in what are known as medical savings accounts. In 1984 Singapore set up a system based on these accounts complemented by public finance for the indigent and catastrophic insurance to cover exceptionally high costs. There are several problems with medical savings accounts. They lose many of the efficiency advantages of risk pooling. Because the money in such accounts is not fungible, they give people with good health an incentive to purchase medical care even when it is unnecessary. And medical savings accounts and catastrophic insurance are most compatible with fee-for-service payment methods, a combination that has consequences for cost escalation and medically inappropriate care. In Singapore per capita health care costs have grown 13 percent a year since 1984, 2 percentage points faster than before the introduction of the system.

Containing costs and improving efficiency

- Increases in health expenditures have not necessarily brought about improved health outcomes.
- Fragmented fee-for-service systems like those in the United States and China, without a single payer or set of rules, have been the least successful in controlling expenditures and guaranteeing access.
- Most OECD countries now have surpluses of physicians, particularly specialists, and of hospital beds. These surpluses are a major factor in cost escalation. For political and other reasons, these surpluses, once created, are hard to reduce, and are best avoided in the first place.
- Cost containment is a major issue in almost all OECD countries, as consumer expectations, new technologies, and aging populations continue to place upward pressure on health expenditures. Empowering consumers and having money follow patients in the context of a global budget appear to be the best strategies for controlling costs. Provider payment approaches such as capitated payments and diagnosis-related groups, if appropriately implemented, appear to control costs without compromising access and quality.

The lessons are clear: Fragmented systems that leave many people uninsured and rely principally on fee-for-service payments are inequitable and inefficient and perform poorly in health outcomes. Many countries are paying close attention to these lessons from the OECD reforms. But China since the 1980s has been moving toward less government support for priority public health activities for all and clinical services for the poor, and greater reliance on a fee-for-service delivery system. If present trends persist, China will move into the 21st century with a poorly performing but nonetheless costly health system.

must finance if they are to be provided at adequate levels. Full funding of priority public health programs is the first priority for public spending on health (table 8.2). This report recommends increasing public funding of the Epidemic Prevention Service over the medium term, from 1.3 billion yuan in 1993 to at least 6.5 billion yuan in 2001. This recommended funding level is based on current cost estimates of fully funding tuberculosis control, expanded immunization programs, endemic disease control, health education, and other important programs and on projected cost increases to 2001. (As income is projected to double in real terms, Epidemic Prevention Service costs will also rise, in large part because of wage increases.)

The government's next priority for health spending should be to subsidize essential health services for the poor, on grounds of poverty assistance and equity. Most of the poor and near poor in China live in rural areas and are concentrated in the 592 poorest counties. To reach the poor most effectively, this report thus recommends geographical targeting (aimed at poor villages in China's poor counties) combined with program targeting (universal subsidies) for a few services that particularly benefit the poor.

TABLE 8.2

Illustrative costs of central and provincial government funding of selected priority health programs

(billions of 1993 yuan)

Health program	1993	2001 (proposed or projected)
Fully fund priority public health programs	1.3[a]	6.5
Subsidize essential health services for the poor		
Subsidies for poor villages in poor counties	—	3.7
Program subsidies targeting specific diseases	—	1.0
Subsidies for urban poor	—	0.5
Promote broader risk pooling		
Subsidies to community financing schemes	—	1.2
Urban health finance initiatives	—	0.1[b]
Total	1.3	13.0
As percentage of government expenditure	0.29	0.93
As percentage of GDP	0.04	0.17
Memo items		
GDP	3,451	7,500
General government expenditures	450	1,400
Proposed incremental tobacco tax (20 percent, ad valorem)	n.a.	10

— Not available.
n.a. Not applicable.
a. Figure refers to actual funding in 1993 for the Epidemic Prevention Service, not full funding.
b. This estimate includes the costs of major studies on pricing and provider payment, monitoring, and evaluation.
Source: For 1993 data, Wei 1996 and World Bank 1996b.

If community financing schemes could be established in poor rural areas, perhaps with technical assistance from the government, the government could channel subsidies for services for the poor through them. China has much experience with these schemes, and they offer many benefits, providing risk pooling for the entire community and efficiencies in the delivery of care. Nonetheless, they are relatively difficult to administer, perhaps especially in poor villages with little administrative capacity.

An alternative or complementary approach would be to provide supply-side subsidies (subsidies to health programs or health facilities in return for providing free care to the poor in the poorest counties). Existing subsidies to township health centers could gradually be redirected to the poorest counties. A reasonable target would be 25 yuan per person now, rising to 50 yuan per person by 2001, in line with income growth. The cost of the subsidies by 2001 would depend on the number of people covered. Based on a rough estimate of about 75 million people living in poor villages in China's 592 designated poor counties, the cost would be 3.75 billion yuan.

Program subsidies for clinical services that disproportionately benefit the poor offer a means for reaching both urban and rural poor. These subsidies might cost 1 billion yuan a year in 2001. Other measures are also needed to improve the urban poor's access to priority health services. The government is trying to reallocate existing resources for the rural and urban poor through such measures as the Three Items Construction Program. But this reallocation is difficult under the current fiscal structure. Better funding for public health programs will help, but the government could also consider redirecting some of the existing general hospital subsidies (4.5 billion yuan) to services for the urban poor. (The Ministry of Health argues that it is committed to providing subsidies for basic salaries of hospital employees and that reallocation is therefore not possible.) These subsidies could be focused on public facilities in poor urban areas, on programs that particularly benefit the urban poor, or, if individual targeting methods can be developed, on poor households. Because there are far fewer urban than rural poor, subsidies to the urban poor would be far smaller, estimated here at 0.5 billion yuan in 2001.

The third priority for government health finance is to support reforms to prices and provider payment mechanisms, and the fourth is to control investments and improve regional planning. These measures could generate sectorwide improvements in efficiency. Illustrative costs for these measures have not been estimated for this report, but they would be much smaller than the costs of the first two priorities.

The fifth priority for public health spending is to promote risk pooling among the broader population in order to provide protection, at a minimum, from catastrophic health expenses. One way to do so would be to provide a small per capita subsidy to rural communities for operating community financing schemes. Such subsidies can be justified on efficiency grounds: the government has an interest in seeing that efficient risk-pooling mechanisms are established in rural areas. This support could be provided in the form of a matching grant to communities that agree to follow certain principles (see chapter 6). The subsidy might be 5 yuan per person now, growing to 10 yuan by 2001 in line with income growth. If the subsidy covered 120 million people in 2001, the total cost to the government would be roughly 1.2 billion yuan.

TABLE 8.3
China's government expenditures in international perspective
(percent)

	Government expenditures as a share of GDP	Central government expenditures as a share of all government expenditures
All countries	39.1	72.3
Industrial countries	47.6	65.9
Developing countries	31.7	77.8
China, 1994		
Budgetary	14.1	40.2
Extrabudgetary	3.8	n.a.

n.a. Not applicable.
Note: Except for China, data are averages over a three-year period ending in 1987 or 1988.
Source: World Bank 1996b.

Government support is also needed to improve the efficiency and coverage of urban insurance systems. And as urban insurance experiments expand throughout China, the government will need to ensure that they are systematically monitored and evaluated to guide medium- and long-term policy choices. The cost of these activities is estimated at 0.1 billion yuan in 2001.

In the long run these health financing policies will yield high returns. They will generate savings and improve health, especially for the poor. They will also pool risks, improve efficiency, and slow health cost inflation while keeping the share of GDP spent on health care relatively low. The needs are clear. The crucial question is: Does China have the resources and will to accomplish these goals in the face of other pressing economic and social issues?

Finding funds for increased public spending on health

Government revenues in China have declined in recent years as a share of GDP. To keep the budget deficit in check, the government reduced its expenditures from 33.8 percent of GDP in 1978 to 17.9 percent in 1994 (79 percent budgetary and 21 percent extrabudgetary). China's budgetary expenditures are far smaller as a share of GDP than those in most other countries. The central government's share of all government expenditures is also unusually small (table 8.3).

TABLE 8.4
Revenue effect of recommended tax measures in China, 2000
(percent)

Tax measure	Incremental revenue as a share of GDP
VAT	2.1
Individual income tax	0.8
Enterprise income tax	1.2
Taxes on pollutants	1.0
Payroll taxes	0.9
Total	6.0

Source: World Bank 1996b.

Several urgent needs call for increased public spending in China: health, education, infrastructure, poverty alleviation, pension reform, environmental protection, and unemployment insurance. The World Bank has recently recommended public finance reforms in China to increase government revenues and enable the government to better address these urgent needs. The tax measures recommended by the World Bank are projected to boost government revenue by an amount equal to 6 percent of GDP (table 8.4).

With these projected increases in government revenue, the proposed priority health programs in table 8.2 should be easily affordable. Public funds recommended for the priority health programs would total only about 13 billion yuan in 2001, or 0.9 percent of projected public spending. Because of fiscal constraints in recent years, however, government officials are reluctant to plan for expanded public spending on health. But if revenues increase less than projected, the programs could be phased in slowly. Moreover, if China increases its tobacco tax, much or all of the additional revenue could be directed to priority health programs, making them that much more affordable.

Glossary

Adverse selection. The tendency of individuals expecting high health expenditures to purchase insurance or purchase a more generous insurance package than people expecting low levels of expenditures.

Basic benefit package. A minimum set, or core, of health services.

Capitation payment. A fixed payment to a provider for each listed or enrolled person served per period of time. Payments will vary according to the number, age, and sex of patients enrolled but not with the number of services rendered per patient.

Community financing. A community-managed scheme whereby a local community group collects and manages funds from households, government, and local industries and organizes the delivery of a package of health benefits for community members. The health risks of community members are pooled, and the benefits include some catastrophic as well as basic care.

Cost containment. A set of steps to control or reduce inefficiencies in the consumption, allocation, or production of health care services that contribute to higher than necessary costs. Inefficiencies in consumption can occur when health services are inappropriately utilized; inefficiencies in allocation exist when a different mix of services could produce greater health benefits; and inefficiencies in production exist when the cost of producing health services could be reduced by using a different combination of resources.

Cost sharing. A provision of health insurance or third-party payment that requires the individual who is covered to pay part of the cost of medical care received. This is distinct from the payment of a health insurance premium, contribution, or tax, which is paid whether medical care is received or not. Cost sharing may be in the form of deductibles, coinsurance, or copayments.

Diagnosis-related groups. Groupings of diagnostic categories drawn from the International Classification of Diseases and modified by the presence of a surgical procedure, patient age, presence or absence of significant comorbidities or complications, and other relevant criteria. Diagnosis-related groups are the case-mix measure used in Medicare's prospective payment system in the United States. Variations on diagnosis-related groups have been adopted in several other countries as a tool for hospital management or for reimbursement.

Disability-adjusted life year. A unit used for measuring both the global burden of disease and the effectiveness of health interventions, as indicated by reductions in the disease burden. It is calculated as the present value of the future years of disability-free life that are lost as a result of the premature deaths or cases of disability occurring in a particular year.

Externalities. Costs or benefits arising from production or consumption that fall on individuals and groups not directly involved in the production or consumption concerned, and which are not compensated for by exchange. For example, immunization of an individual against an infectious disease can block the transmission of the disease to other individuals who are not directly involved and who pay nothing for the protection they receive. This creates an external benefit. And one person's inappropriate use of an antibiotic will frequently hasten spread of resistance to it, creating external costs borne by others.

Fee-for-service. Payments to a provider for each item or service rendered.

Global budget. An aggregate cash sum, fixed in advance, intended to cover the total cost of a service, usually for one year ahead.

Health insurance. Financial protection against the medical care costs arising from disease or accidental bodily injury. Such insurance usually covers all or part of the medical costs of treating the disease or injury. Insurance may be obtained on either an individual or a group basis.

Health maintenance organization. An organization that accepts responsibility for organizing and providing a defined set of health services for its enrolled population, in exchange for a predetermined, fixed, periodic payment for each person or family unit enrolled. The payment is fixed without regard to the amount of actual services provided to an enrollee.

Health planning. Planning concerned with improving health, whether undertaken comprehensively for a whole community or for a particular population, type of health service, institution, or health program. The components of health planning include data assembly and analysis, goal determination, action recommendation, and implementation strategy.

Managed competition. Government regulation of health insurance and health care markets using competition as the means to achieve efficiency objectives within a framework of government intervention.

Moral hazard. In health insurance, this refers to the fact that people who are insured tend to act in ways that increase health expenditures. They may take fewer precautions against avoidable events, or, more important, patients and providers may increase the use of services because insurance subsidizes prices at the point of service.

Out-of-pocket payments. Payments made directly by a patient without reimbursement by any insurance. They include cost sharing.

Prepayment. Usually refers to any payment to an organization for anticipated services (such as an expectant mother paying in advance for maternity care). Prepayment is distinguished from insurance because it involves payment to organizations that, unlike an insurance company, take responsibility for arranging and providing needed services as well as paying for them (such as health maintenance organizations, prepaid group practices, and medical foundations).

Public good. A good or service whose benefits may be provided to a group at no more cost than that required to provide it for one person. The benefits of the good are indivisible and individuals cannot be excluded. For example, a public health measure that eradicates smallpox protects all, not just those paying for the vaccination.

Public health. The science dealing with the protection and improvement of the population's health by organized community effort. Public health activities are generally those that are less amenable to being undertaken by individuals or that are less effective when undertaken on an individual basis. They do not typically include direct personal health services. Public health activities include immunization; sanitation; preventive medicine, quarantine, and other disease control activities; occupational health and safety programs;

assurance of the healthfulness of air, water, and food; health education; and epidemiology.

Risk pooling. The process of distributing the probability of financial loss across multiple parties. Health insurance, for example, is a way of distributing the uneven burden of medical expenditures across a subgroup of the population.

Salary payment. Remuneration that is fixed for a period of time and does not vary either with the number of individuals served or with the number of services rendered, although rate of salary change can depend on performance.

Selection bias. The tendency for multiple health plans or providers to attract, intentionally or unintentionally, an uneven (biased) distribution of health risks. Also known as risk selection. When done intentionally by insurance companies to discourage participants likely to need substantial care, the practice is sometimes known as "cherry-picking."

Social health insurance. A term mainly used to denote compulsory, or public, health insurance, usually part of a social security system, which is funded from specific (mainly payroll) contributions and managed by a government agency or autonomous organization such as a sickness fund, mutual aid society, or private insurer.

Supplier-induced demand. The ability of providers to use their authority to boost demand for their services, for the services of colleagues, or for the sale of drugs or tests in which they have a financial stake, above the level that would be demanded if patients had adequate and accurate information on medical benefits and costs.

Third-party payer. Any organization, public or private, that pays or insures health care expenses for beneficiaries at the time they are patients. The first party is the patient and the second party is the provider. Third parties may be private insurers, quasi-public bodies such as sickness funds, and government bodies.

Voluntary health insurance. Health insurance that is taken up and paid for at the discretion of individuals, or employers on behalf of individuals. Voluntary insurance can be offered by a private, public, or quasi-public body.

Yellow book. Generic term in China for hospital and health service price lists for the thousands of products and services sold.

Note

This glossary is based on OECD (1994, pp. 9–11) and Alpha Center (n.d.).

Annex

TABLE A.1
Gross domestic product, price deflators, and nominal exchange rate, 1978–94

Year	GDP (billions of yuan) Current prices	GDP (billions of yuan) Constant 1993 prices	GDP price deflator index 1990=1	GDP price deflator index 1993=1	Nominal official exchange rate (yuan to U.S. dollar; annual average)
1978	362.4	880.9	0.541	0.411	1.7
1979	403.8	948.2	0.560	0.426	1.6
1980	451.8	1,020.8	0.582	0.443	1.5
1981	486.2	1,067.4	0.599	0.456	1.7
1982	529.5	1,158.6	0.601	0.457	1.9
1983	593.5	1,275.2	0.612	0.465	2.0
1984	717.1	1,471.1	0.641	0.487	2.3
1985	896.4	1,669.6	0.706	0.537	2.9
1986	1,020.2	1,815.4	0.739	0.562	3.4
1987	1,196.3	2,027.2	0.776	0.590	3.7
1988	1,492.8	2,256.4	0.870	0.662	3.7
1989	1,690.9	2,348.0	0.947	0.720	3.8
1990	1,853.1	2,436.8	1.000	0.760	4.8
1991	2,161.8	2,664.3	1.067	0.811	5.3
1992	2,663.5	3,040.4	1.152	0.876	5.5
1993	3,451.5	3,451.5	1.315	1.000	5.8
1994	4,500.6	3,860.6	1.533	1.166	8.6

Source: China, State Statistical Bureau 1995, p. 32.

TABLE A.2A

Gross domestic product, population, and health spending, 1978–93

(constant 1993 prices)

Year	GDP (billions of yuan)	Year-end population (tens of thousands)	GDP per capita (yuan)	Total health spending (billions of yuan)	Health spending per capita (yuan)
1978	880.9	96,259	915	25.8	27
1979	948.2	97,542	972	28.7	29
1980	1,020.8	98,705	1,034	31.6	32
1981	1,067.4	100,072	1,067	34.2	34
1982	1,158.6	101,541	1,141	39.0	38
1983	1,275.2	103,008	1,238	43.0	42
1984	1,471.1	104,357	1,410	47.9	46
1985	1,669.6	105,851	1,577	49.0	46
1986	1,815.4	107,507	1,689	54.6	51
1987	2,027.2	109,300	1,855	62.5	57
1988	2,256.4	111,026	2,032	71.1	64
1989	2,348.0	112,704	2,083	77.8	69
1990	2,436.8	114,333	2,131	84.8	74
1991	2,664.3	115,823	2,300	94.2	81
1992	3,040.4	117,172	2,595	103.3	88
1993	3,451.5	118,517	2,912	132.1	111

Source: China, State Statistical Bureau 1986 (p. 71), 1994 (p. 59), 1995 (p. 32); Wei 1995.

TABLE A.2B

Annual growth in population and in per capita GDP and health spending, 1978–93

(percent)

	1978–86	1986–93	1978–93
Population	1.4	1.4	1.4
GDP per capita	7.7	7.8	7.7
Health spending per capita	8.0	11.2	9.5

Source: China, State Statistical Bureau 1986 (p. 71), 1994 (p. 59), 1995 (p. 32); Wei 1995.

TABLE A.3

National health expenditure, 1978, 1986, and 1993

(millions of 1993 yuan)

Funding source	1978	1986	1993
Government budget[a]	7,292 (28)	17,288 (32)	18,878 (14)
Government and state enterprise insurance systems	7,689 (30)	18,274 (33)	46,108 (36)
Out-of-pocket payments	5,268 (20)	14,185 (26)	56,106[b] (42)
Rural cooperative medical system	5,109 (20)	2,918 (5)	2,243 (2)
Other	428 (2)	1,956 (4)	8,713 (6)
Total	25,786 (100)	54,621 (100)	132,048 (100)
Memo item			
Total health expenditure as percentage of GDP	2.9	3.2	3.8

Note: Figures in parentheses are percentage shares of the total.

a. Excludes the government insurance system.

b. Reflects the Health Economics Institute's reestimate of out-of-pocket payments in March 1996.

Source: China, Ministry of Health data (provided in May 1995).

TABLE A.4

Government spending on health, selected years, 1978–93

(millions of 1993 yuan)

Spending category	1978	1986	1990	1991	1992	1993
Recurrent health budget	5,294	10,598	10,686	10,658	10,965	10,786
Traditional Chinese medicine	0	643	869	901	951	917
Family planning fund	0	1,409	1,730	1,965	2,211	2,292
Research	160	197	132	134	177	251
Higher education	0	0	0	899	857	899
Capital investment	628	2,107	1,400	895	877	1,144
Other ministries	1,209	2,335	2,113	1,226	1,406	1,415
Subtotal	7,291	17,288	16,930	16,680	17,443	18,878
Government insurance system	1,178	3,360	5,834	6,234	6,667	11,667
Total	8,469	20,648	22,764	22,914	24,110	25,323

Source: China, Ministry of Health data (provided in July 1995).

TABLE A.5

Components of recurrent health budget, selected years, 1978–93

(millions of 1993 yuan)

Spending category	1978	1986	1990	1991	1992	1993
Hospital operating expenses	1,794	4,488	4,254	4,072	4,290	4,183
Subsidies for health centers	1,421	1,903	1,969	2,040	2,233	2,145
Epidemic prevention fund	944	1,544	1,583	1,655	1,948	1,305[a]
Maternal and child care fund	0	295	402	429	536	324[a]
Pharmaceutical control fund	0	144	147	164	194	229
Professional middle school fund	264	511	503	517	590	826
Training fund	0	100	0	108	142	156
Rural cooperative medical system fund	89	50	39	38	33	27
Kindergarten fund	0	7	0	7	9	9
Indigent patients' hospital fund	0	18	0	17	18	15
Other	783	1,662	1,789	1,610	1,856	1,919
Total recurrent health budget	5,295	10,722	10,685	10,658	11,850	11,139

a. Before 1990 government budget data excluded revenues from user fees; from 1991 onward the data included user fees. The budget data for the epidemic prevention and maternal and child care funds for 1993 were revised to exclude net user fees, but 1991 and 1992 figures have not yet been revised. Discrepancies between tables A.4 and A.5 in the total recurrent budget figures arise largely as a result of this inconsistency in accounting for sources of funds.

Source: China, Ministry of Health data (provided May 1995); Health Economics Institute March 1996 revised estimates for epidemic prevention and maternal and child care funds for 1993.

TABLE A.6

Health insurance coverage and health spending of the rural population, 1993

Group	Population (millions)	Income per capita (1993 yuan)	Percentage insured	Health spending per capita (1993 yuan)	Total health spending (billions of 1993 yuan)
Rural population	900	750	10	60	54
Top quartile	225	920	40	98	22
Third quartile	225	668	0	67	15
Second quartile	225	489	0	44	10
Bottom quartile	225	361	0	31	7

Source: Estimates based on data from Wei 1995; the Health Economics Institute's January 1996 revised estimates of national health expenditures; Zhao Zhuyan and Lusheng Wang 1995; China, State Statistical Bureau 1994. Health Economics Institute staff adjusted the 1993 National Health Services Survey data to estimate spending by subgroups.

TABLE A.7

Revenues and government subsidies of health institutions providing hospital services, 1993

(billions of 1993 yuan)

Source of revenue	Hospitals	Township health centers	Total
Sales of goods and services	44.8	12.4	57.2
Medical treatment	18.0	3.2	21.2
Pharmaceuticals and others	26.8	9.2	36.0
Government subsidies	4.6	2.0	6.6
Total	49.4	14.3	63.8

Source: Health Economics Institute estimates, based on China, Ministry of Health 1994.

TABLE A.8

Average number of hospital beds and health care personnel per hospital, 1993 and 1994

Type of hospital	1993			1994		
	Beds	Personnel	Medical and technical personnel	Beds	Personnel	Medical and technical personnel
Hospitals at county level and above	137	189	145	139	192	148
Rural township hospitals	16	21	18	114	120	17
Other	45	59	47	42	56	44

Source: China, State Statistical Bureau 1995, p. 667.

TABLE A.9

Utilization of beds in hospitals at county level and above, selected years, 1985–94

Indicator	1985	1989	1990	1992	1993	1994
Government hospitals						
Annual bed turnover	19.9	19.9	19.9	18.9	17.9	17.9
Bed days in use	320.9	315.9	313.9	303.9	267.9	263.9
Bed occupancy rate (percent)	87.9	86.9	85.9	83.9	75.9	72.9
Average length of stay (days)	15.9	15.9	15.9	15.9	15.9	14.9
State enterprise hospitals						
Annual bed turnover	14.9	14.9	13.9	12.9	12.9	12.9
Bed days in use	253.9	259.9	255.9	245.9	221.9	226.9
Bed occupancy rate (percent)	69.9	70.9	69.9	67.9	60.9	62.9
Average length of stay (days)	16.9	16.9	17.9	17.9	16.9	16.9
All hospitals						
Annual bed turnover	18.9	18.9	17.9	16.9	15.9	15.9
Bed days in use	302.9	299.9	296.9	286.9	259.9	251.9
Bed occupancy rate (percent)	82.9	81.9	80.9	78.9	71.9	69.9
Average length of stay (days)	15.9	15.9	15.9	16.9	15.9	15.9

Source: China, State Statistical Bureau 1995, p. 671.

References

Background papers and consultant reports

Akin, John S., and Paul Hutchinson. 1995. "Health Insurance and the Rural Poor in China." World Bank, Human Development Department, Washington, D.C.

Berman, Peter, Gilles Fortin, Vernon Hicks, William McGreevey, and J. Brad Schwartz. 1995. "China's National Health Accounts." World Bank, Human Development Department, Washington, D.C.

Cai, Renhua. 1995b. "Trial of Reforming the Employee Medical Care Insurance System in Zhenjiang." Paper presented at health care seminar, Beijing, 2–5 May. China, Ministry of Health, Beijing.

Chen, Xiaoming. 1995. "The Analysis of Hospitals/Doctors Prescribing Behavior." Paper presented at health care seminar, Beijing, 2–5 May. Shanghai Medical University, Shanghai.

Cross, Harry E. 1995. "Health Care and Poverty in China: Issues in Providing Basic Services to the Poor." World Bank, Human Development Department, Washington, D.C.

Gu, Xingyuan, Baogang Shu, Bo Cha, Jie Yu, and Wei Deng. 1996. "Chinese Mortality and Life Expectancy." Shanghai Medical University, School of Public Health, Department of Health Statistics and Social Medicine, Shanghai.

Hammer, Jeffrey S. 1996. "Setting the Context of Health Care Finance in China." World Bank, Policy Research Department, Washington, D.C.

Hill, Kenneth, and Akiko Maeda. 1997. "Child Mortality Trends in China." World Bank, Human Development Department, Washington, D.C.

Hou, Yan, and Heyu Zhou. 1995. "The Relationship Between Health Capital Investment and Health Operating Expenses." Paper presented at health care seminar, Beijing, 2–5 May. China, State Planning Commission, Beijing.

Hsiao, William C. 1996. "Notes on Health Reform in China." Harvard University, Cambridge, Mass.

Hu, Haobo. 1995. "Report about High-Tech Medical Equipment (HTME)." Paper presented at health care seminar, Beijing, 2–5 May.

Hu, Shanlian. 1995. "Pricing Policy of Health Services in China." Paper presented at health care seminar, Beijing, 2–5 May.

Hu, Teh-Wei. 1995. "Cigarette Taxation in China: Lessons from International Experience." Paper prepared for the World Bank, Human Development Department, Washington, D.C.

Jamison, Dean. 1996. "Outside Influences on Health Status in China." World Bank, Human Development Department, Washington, D.C.

Jin, Shuigao. 1995a. "Case Study on Health Finance of Shanxi, Jiangsu, and Guizhou Provinces." Paper presented at health care seminar, Beijing, 2–5 May.

———. 1995c. "Trends in the Infant Mortality Rate and Life Expectancy Change, 1949–1955."

———. 1995d. "Trends in the Prevalence and Mortality for Tuberculosis and Schistosomiases, 1980–1994."

Luo, Wujin. 1995. "The Health Financing Study in Rural Areas of China." Paper presented at health care seminar, Beijing, 2–5 May.

Mao, Zhenzhong. 1995. "Studies in Sichuan Rural Health Insurance Experiments." Paper presented at health care seminar, Beijing, 2–5 May. West China Medical University, Chengdu, Sichuan.

Meng, Jianguo. 1994. "Pilot Reform of the Separation Between State-Owned Large Enterprises and Their Hospitals." Paper presented at health care seminar, Beijing, 2–5 May. China, Ministry of Health, Beijing.

———. 1995. "Financial Analysis of Medical Facilities." Paper presented at health care seminar, Beijing, 2–5 May. China, Ministry of Health, Beijing.

Wei, Ying. 1995. "An Introduction to Health Financing Patterns in China." National Health Economics Institute, Beijing.

Yang, Hui. 1995. "Political Research on Re-establishing and Improving the Chinese Rural Cooperative Medical Care System." Paper presented at health care seminar, Beijing, 2–5 May. Beijing Medical University, Beijing.

Yip, Winnie. 1995. "The Role of Government in Health Care Financing in China." Consultant Report to the China Health Care Financing Mission, 10 April–6 May. World Bank, Human Development Department, Washington, D.C.

———. 1996. "Notes on the Zhenjiang Demonstration Project." Harvard University, Cambridge, Mass.

Yu, Dezhi, and Xinming Liu. 1995. "China's Economic Reform in the 1980s and Its Impact on Health Care Services." Paper presented at health care seminar, Beijing, 2–5 May. China, Ministry of Health, Beijing.

Zhao,Yuxin. 1995. "Policy Issues on Governmental Funding to Health Work in the People's Republic of China." National Health Economics Institute, Beijing.

Zhao, Zhuyan, and Lusheng Wang. 1994. "Present Ways to Raise Funds for Health Care in the Countryside of China." Paper presented at health care seminar, Beijing, 2–5 May. National Health Economics Institute, Beijing.

Zhenjiang, City of. 1995. "Progress Report on the Pilot Health Insurance Reform." Jiangsu Province.

Bibliography

Akin, John S., and Paul Hutchinson. 1995. "Health Insurance and the Rural Poor in China." World Bank, Human Development Department, Washington, D.C.

Alpha Center. n.d. *Glossary of Terms as Commonly Used in Health Care.* Washington, D.C.

Barnum, Howard, Joseph Kutzin, and Helen Saxenian. 1994. "Incentives and Provider Payment Methods." World Bank, Population, Health, and Nutrition Department, Washington, D.C.

Berk, M.L., and A.C. Monheit. 1992. "The Concentration of Health Expenditures: An Update." *Health Affairs* 11(4): 145–49.

Berman, Peter, Gilles Fortin, Vernon Hicks, William McGreevey, and J. Brad Schwartz. 1995. "China's National Health Accounts." World Bank, Human Development Department, Washington, D.C.

Bloom, Gerald, Xingyuan Gu, and Shenglan Tang. 1995. "Financing Rural Health Services in China in the Context of Economic Reform." *Journal of International Development* 7(3): 423–41.

Cai, Renhua. 1995a. "An Exploratory Discussion of Chinese Workers Medical Insurance Reform." Paper presented at International Health Policy Program Sixth Meeting of Participants, Beijing, 2–7 October. China, Ministry of Health, Legal Affairs Department, Beijing.

———. 1995b. "Trial of Reforming the Employee Medical Care Insurance System in Zhenjiang." Paper presented at health care seminar, Beijing, 2–5 May. China, Ministry of Health, Legal Affairs Department, Beijing.

Cardoso, Eliana. 1995. "Can China Rescue Its State-Owned Enterprises?" World Bank, East Asia and Pacific, Country Departments II and III, Washington, D.C.

Chang, Ying, Fengying Zhai, Wenjun Li, Keyou Ge, Daxun Jin, and M. de Onis. 1994. "Nutritional Status of Preschool Children in Poor Rural Areas of China." *Bulletin of the World Health Organization* 72(1): 105–12.

Chen, Hu Lin. 1993. "Cooperative Medical System in China." *International Journal of Health Services* 23(4): 733–42.

Chen, Jie. 1994. "Evaluation of Technology Cost and Diffusion and Policy Implications." *Health Social Science,* III, pp. 4–5.

Chen, Shaohua, and Martin Ravallion. 1996. "Data in Transition: Assessing Rural Living Standards in Southern China." World Bank, Policy Research Department, Washington, D.C.

Chen, Xiaoming. 1994. "Importance of Drug Income to Cross-Subsidize Hospitals' Losses." *Journal of Health Resources* 2:2–30.

———. 1995. "The Analysis of Hospitals/Doctors Prescribing Behavior." Paper presented at health care seminar, Beijing, 2–5 May. Shanghai Medical University, Shanghai.

China, Government of. 1993. "China Social Sector Policies. Extracts from the Decision of the CPC Central Committee on Issues Concerning the Establishment of a Socialist Market Economic Structure." Translated. 14 November.

China, Ministry of Health. 1991. *Selected Edition on Health Statistics of China, 1978–1990.* Center for Health Statistics Information, Beijing.

———. 1993a. "Background Papers for the International Seminar on Health Reform under Socialist Market Economy." Department of Planning and Finance, Beijing.

———. 1993b. *Medical and Health Services in China.* Edited by the Foreign Affairs Bureau of the Ministry of Health. Beijing.

———. 1993c. "Proposal for WHO Support for a Field Study on Rural Health Insurance Systems in the People's Republic of China, 1994–95." Beijing.

———. 1994. *Statement of Health Budget and Expenditure.* Beijing.

———. 1995a. "Health-Care Finance, 1978–1993." Tables. Department of Planning and Finance, Beijing.

———. 1995b. *National Strategic Plan for the Prevention of AIDS and STD in the People's Republic of China, 1996–2000.* Beijing.

China Network for Training and Research in Health Economics and Financing. 1994a. "Financing and Organization of Health Care for the Poor Rural Population in China: Report on Project Activities." Paper presented at HIP Asia Meeting, Olongapo City, Philippines, 16–17 May.

———. 1994b. "Paper Catalogue of Financing and Distributing in Medical Care and Health Finance in China."

———. 1995. "Summary of the State Government Policy Forum on Health Care Financing and Organization in Poor Rural Areas of China." Health Economics Institute, Beijing.

China Network for Training and Research in Health Economics and Financing and Harvard School of Public Health. 1995. "Study of Health Care Financing and Organization in Impoverished Rural Areas of China." Paper presented at International Seminar on Financing and Organization of Health Care for Poor Rural Population in China, Fragrant Hills Hotel, Beijing, 8–10 October.

———. 1996. "Report to UNICEF and International Health Policy Program on the Study of Thirty Poor Counties."

China, State Pharmaceutical Administration. 1995. "Management of China's Pharmaceutical Market and Its Problems." Paper presented at International Federation of Pharmaceutical Manufacturers Association conference, Providing Quality Medicines for Health Care, Bangkok, 6–8 December.

China, State Statistical Bureau. 1985. *Statistical Yearbook of China.* Beijing: China Statistical Publishing House.

———. 1986. *Statistical Yearbook of China.* Beijing: China Statistical Publishing House.

———. 1992. *Statistical Yearbook of China.* Beijing: China Statistical Publishing House.

———. 1993. *Statistical Yearbook of China.* Beijing: China Statistical Publishing House.

———. 1994. *Statistical Yearbook of China.* Beijing: China Statistical Publishing House.

———. 1995. *Statistical Yearbook of China.* Beijing: China Statistical Publishing House.

China Tuberculosis Control Collaboration. 1996. "Results of directly observed short-course chemotherapy in 112,842 Chinese patients with smear-positive tuberculosis." *Lancet* 347 (10 February): 358–62.

Chinese Journal of Hospital Management. 1993. No. 9.

Chu, David K.W. 1994. "Economic Development and the Health Care System in Hong Kong." *Health Policy* 28: 211–34.

Clayton, S., H. Yang, J. Guan, Z. Lin, and R. Wang. 1993. "Hepatitis B Control in China: Knowledge and Practices among Village Doctors." *American Journal of Public Health* 83: 1685–88.

Cretin, S., N.H. Duan, A.P. Williams Jr., X.Y. Gu, and Y.Q. Shi. 1990. "Modeling the Effect of Insurance on Health Expenditures in the People's Republic of China." *Health Services Research* 25(4): 667–85.

Cretin, S., Albert P. Williams, Jeffrey J. Sine, Dechun Luo, Dexiao Zhang, Yuanqiu Shi, Shuqin Yang, and Tingwei He. 1995. "China Rural Health Insurance Experiment." Paper presented at International Health Policy Program Sixth Meeting of Participants, Beijing, 2–7 October.

Cross, Harry E. 1995. "Health Care and Poverty in China: Issues in Providing Basic Services to the Poor." World Bank, Human Development Department, Washington, D.C.

Dai, Zhicheng. 1995. "China's Public Health Work: Achievements and Prospects." Paper presented at International Health Policy Program Sixth Meeting of Participants, Beijing, 2–7 October.

De Geyndt, Willy. 1994. "Managing the Quality of Health Care in Developing Countries." World Bank, Asia Technical Department, Human Resources and Social Development Division, Washington, D.C.

De Geyndt, Willy, X. Zhao, and L. Shunli. 1992. *From Barefoot Doctor to Village Doctor in Rural China.* World Bank Technical Paper 187. Washington, D.C.

Feng, Xueshan, Shenglan Tang, Gerald Bloom, Malcolm Segall, and Xingyuan Gu. 1995. "Cooperative Medical Schemes in Contemporary Rural China." *Social Science and Medicine* 41 (8): 1111–18.

Food and Agriculture Organization. 1994. *FAO Annual Yearbook.* Rome.

Fu, Wei, and others (China Network of Training and Research in Health Economics and Financing). 1995. "Chinese Poor Rural Areas Demand-Side Health Financing Systems Research."

Greenhalgh, Susan. 1994. "The Social Dynamics of Child Mortality in Village Shaanxi." Population Council Research Division Working Paper 66. New York.

Griffin, Charles C. 1992. *Health Care in Asia: A Comparative Study of Cost and Financing.* World Bank Regional and Sectoral Studies. Washington, D.C.

Gu, Xingyuan. 1992. "Infant Mortality Rate in 8 Provinces in China." *Chinese Journal of Health Statistics* 9(3): 7.

———. 1995. "Impacts of Research on Policy: Financing of Health Care in Poor Rural China." Paper presented at International Health Policy Program Sixth Meeting of Participants, Beijing, 2–7 October.

Gu, Xingyuan, Gerald Bloom, Shenglan Tang, Yingya Zhu, Shouqi Zhou, and Xingbao Chen. 1993. "Financing Health Care in Rural China: Preliminary Report of a Nationwide Study." *Social Science and Medicine* 36(4): 385–91.

Gu, Xingyuan, Baogang Shu, Bo Cha, Jie Yu, and Wei Deng. 1996. "Chinese Mortality and Life Expectancy." Shanghai Medical University, Shanghai.

Gu, Xingyuan, and Shenglan Tang. 1995. "Reform of the Chinese Health Care Financing System." *Health Policy* 32: 181–91.

Gu, Xingyuan, Shenglan Tang, Gerald Bloom, and Henry Lucas. 1995a. "Health Expenditures and Finance in Three Poor Counties of China." Institute of Development Studies Working Paper 21. University of Sussex, Brighton, England.

———. 1995b. "A Strategy for Health Sector Reform in Rural China." *Chinese Primary Health Care* 9(8): 1–49.

Gu, Xingyuan, Shenglan Tang, and Suhua Cao. 1995. "Financing of Health Services in Poor Rural China: A Case Study in Donglan County." Shanghai Medical University, School of Public Health, Department of Health Statistics and Social Medicine, Shanghai.

Gu, Xingyuan, Kaining Zhang, and J. Kaufman. 1993. "A Proposal to Investigate the Financing, Provision, and Utilization of Reproductive Health Services in Rural China (July 1993–July 1996)." Shanghai Medical University, School of Public Health, Department of Health Statistics and Social Medicine, Shanghai.

Guiyang People's Provincial Hospital. 1995. "Revenue and Expenditure Tables for 1994." Provided to World Bank mission in Guiyang, 24 April.

Guizhou Health Bureau, Provincial Health and Anti-Epidemic Preventive Station. 1995. "The Development of Guizhou

Provincial Health and Anti-Epidemic Preventive Station and the Existing Difficulties." Paper presented to World Bank mission in Guiyang, 24 April.

Guo, Shuqing. 1994. "Present Situation, Experiments and Prospects of China's Social Insurance System Reform." Paper presented at International Monetary Fund Institutional Seminar on Public Sector Pension Reform, Washington, D.C., 8–12 August. China, State Commission for Restructuring the Economic System (SCRES), Department of Comprehensive Programming and Experimenting.

Hammer, Jeffrey S. 1996. "Setting the Context of Health Care Finance in China." World Bank, Policy Research Department, Washington, D.C.

Henderson, Gail, J. Akin, Zhiming Li, Shuigao Jin, Haijiang Ma, and Keyou Ge. 1992. "Equity and Utilization of Health Services: Report of an Eight-Province Survey in China." Paper presented at the Annual Meeting of the Association for Asian Studies, Washington, D.C., April.

Henderson, Gail, M. Goldman, and D. Simon. 1988. "Issues in the Modernization of Medicine in China." *Science and Technology in China*. Cambridge, Mass.: Harvard University Press.

Henderson, Gail, Shuigao Jin, John Akin, Zhiming Li, Jianmin Wang, Haijiang Ma, Yunan He, Xiping Zhang, Ying Chang, and Keyou Ge. 1995. "Distribution of Medical Insurance in China." *Social Science and Medicine* 41(8): 1119–30.

Hill, Kenneth. 1988. *Demographic Trends in China from 1950 to 1982*. World Bank Discussion Paper 22. Washington, D.C.

Hill, Kenneth, and Akiko Maeda. 1997. "Child Mortality Trends in China." World Bank, Human Development Department, Washington, D.C.

Hill, Kenneth, and Rohini Pande. Forthcoming. *Trends in Child Mortality in the Developing World, 1960 to 1995*. New York: UNICEF.

Hong, Phua Kai. 1991. "Privatization and Restructuring of Health Services in Singapore." Institute of Policy Studies Occasional Paper 5. Singapore.

———. 1993. "Health Cost Containment Policies in Singapore." National University of Singapore, Department of Community, Occupational, and Family Medicine, Singapore.

Hong, Wang. 1995. "Analysis of Demand in Poor Rural China: Implications for Financing and Organizing Health Care." Paper presented at International Seminar on Financing and Organization of Health Care for Poor Rural Population in China, Fragrant Hills Hotel, Beijing, 8–10 October.

Hou, Yan, and Heyu Zhou. 1995. "The Relationship Between Health Capital Investment and Health Operating Expenses." Paper presented at health care seminar in Beijing, 2–5 May. China, State Planning Commission, Beijing.

Hsiao, William C. 1984. "Transformation of Health Care in China." *New England Journal of Medicine* 310(14): 932–36.

_______. 1995a. "The Chinese Health Care System: Lessons for Other Nations." *Social Science and Medicine* 41(8): 1047–55.

_______. 1995b. "Community Health Care." Paper presented at International Seminar on Financing and Organization of Health Care for Poor Rural Population in China, Fragrant Hills Hotel, Beijing, 8–10 October.

______. 1996. "Notes on Health Reform in China." Harvard University, Cambridge, Mass.

Hsiao, William C., Yuanli Liu, and others. 1996. "Provision and Financing of Health Care in China's Poverty Regions." China Monograph. UNICEF, New York.

Hu, Haobo. 1995. "Report about High-Tech Medical Equipment (HTME)." Paper presented at health care seminar, Beijing, 2–5 May.

Hu, Shanlian. 1995a. "Financing and Cost of 15 Selected Items of MCH Services in the Poor Counties of China." Paper presented at International Health Policy Program Sixth Meeting of Participants, Beijing, 2–7 October.

———. 1995b. "Policy Impact Exercise Disseminating Research Findings on Policy: Financing and Organization of Health Care for Poor Rural Population in China." Paper presented at International Health Policy Program Sixth Meeting of Participants, Beijing, 2–7 October.

———. 1995c. "Pricing Policy of Health Services in China." Paper presented at health care seminar, Beijing, 2–5 May.

———. 1996. "China Health Sectoral Report: Problems and Prospects." Final meeting on Social Sector Issues in Asian Transition Economies, Asian Development Bank, Manila.

Hu, Shanlian, Songtao Liu, Yan Ma, Yunjie Song, Ming Wu, Jialin Jiang, and Leiya Han (China Network of Training and Research in Health Economics and Financing). 1995. "Financing and Cost of 15 Selected Items of MCH Services in the Poor Counties of China." Paper presented at the International Health Policy Program Research, Writing, and Dissemination Workshop, Washington, D.C., 1–9 March.

Hu, Teh-Wei. 1988. "Diffusion of Western Medical Technology in China since the Economic Reform." *International Journal of Technology Assessment in Health Care* 4: 345–58.

———. 1995. "Cigarette Taxation in China: Lessons from International Experience." Paper prepared for the World Bank, Human Development Department, Washington, D.C.

———. n.d. "Medical Technology Transfer in Major Chinese Teaching Hospitals with Special Reference to Diagnostic Equipment." University of California, School of Public Health, Berkeley.

Hu, Teh-Wei, Jushan Bai, Theodore E. Keeler, Paul G. Barnett, and Hai-Yen Sung. 1994. "The Impact of California Proposition 99, a Major Anti-Smoking Law, on Cigarette Consumption." *Journal of Public Health Policy* 15(1): 26–36.

Hu, Teh-Wei, Theodore E. Keeler, Hai-Yen Sung, and Paul G. Barnett. 1995. "The Impact of California Anti-Smoking Legislation on Cigarette Sales, Consumption, and Prices." *Tobacco Control* 4(supplement 1): S34–38.

Hu, Teh-Wei, Elizabeth Li, and Zhi Hua Lin. 1995. "Insurance Coverage and Out-of-Pocket Medical Expenditures among Chinese Urban Workers." University of California, School of Public Health, Berkeley.

Hu, Teh-Wei, Qiufang Ren, Theodore E. Keeler, and Joan Bartlett. 1995. "The Demand for Cigarettes in California and Behavioral Risk Factors." *Health Economics* 4: 7–14.

Hu, Teh-Wei, Hai-Yen Sung, and Theodore E. Keeler. 1995. "Reducing Cigarette Consumption in California: Tobacco Taxes vs an Anti-Smoking Media Campaign." *American Journal of Public Health* 85(9): 1218–22.

Hu, Zu-liu. 1994. "Social Protection, Labor Market Rigidity, and Enterprise Restructuring in China." IMF Paper on Policy Analysis and Assessment. International Monetary Fund, Fiscal Affairs Department, Washington, D.C.

Jalan, Jyotsna, and Martin Ravallion. 1996. "Transient Poverty in Rural China." World Bank, Policy Research Department, Washington, D.C.

Jamison, Dean T. 1985. "China's Health Care System: Policies, Organization, Inputs and Finance." In Scott B. Halstead, Julia

A. Walsh, and Kenneth Warren, eds., *Good Health at Low Cost*. New York: Rockefeller Foundation.
———. 1996. "Outside Influences on Health Status in China." World Bank, Human Development Department, Washington, D.C.
Jiang, Z., E. Zhang, S. Chen, and H. Yu. 1995. *Statistics of 1992 Fertility Sampling Survey in China*. Beijing: China Population Publishing House.
Jin, Shuigao. 1995a. "Case Study on Health Finance of Shanxi, Jiangsu, and Guizhou Provinces." Paper presented at health care seminar, Beijing, 2–5 May.
———. 1995b. "An Evaluation on Smoking-Induced Health Costs in China (1988–1989)." *Biomedical and Environmental Sciences* 8: 342–49.
———. 1995c. "Trends in the Infant Mortality Rate and Life Expectancy Change, 1949–1955."
———. 1995d. "Trends in the Prevalence and Mortality for Tuberculosis and Schistosomiases, 1980–1994."
Jin, Shuigao, and others. 1994. "Preliminary Analysis on the Impact of Economic Development on Health Status in Rural Areas." In Chunming Chen and Zongming Shao, eds., *Food, Nutrition and Health Status of Chinese in 7 Provinces, 1990, China*. Beijing: China Statistical Publishing House.
Kaji, G. 1993. "Health Challenges for China." Draft speech for seminar commemorating ten years of operational collaboration between China and the World Bank in the health sector. World Bank, China and Mongolia Department, Washington, D.C.
Kan, Xuegui. 1990. "Village Health Workers in China: Reappraising the Current Situation." *Health Policy and Planning* 5(1): 40–48.
Keidel, Albert. 1995. "China: Regional Disparities." World Bank, China and Mongolia Department, Washington, D.C.
Kessel, R. 1958. "Price Discrimination in Medicine." *Journal of Law and Economics* 1: 1–19.
Khoman, Sirilaksana. 1995. "Rural Health Financing: Thailand's Experience." Paper presented at International Health Policy Program International Seminar, Beijing, 8–11 October.
Krieg, Renate, and Monika Schadler, eds. 1994. *Social Security in the People's Republic of China*. Workshop on Social Security in the PRC, 5–7 February 1993. Hamburg.
Li, Heng (Betsy Heng Li). 1993. "A Study of Current Government Health Insurance System and Its Expenditure in China." Chinese Medical Association, Beijing.
Li, Jiali. 1994. "China's Family Planning Program: How, and How Well, Did It Work?" Population Council Research Division Working Paper 65. New York.
Li, Zhiming. 1994. "Employment-Based State Medical Insurance and Job Selection in China: A Switching Simultaneous Equation Model." PhD diss., University of North Carolina, Chapel Hill, Department of Health Policy Administration.
Libo County Hospital. 1995. "Report to the World Bank Team on County Hospital Financial Operations." Paper prepared for World Bank mission and presented in Libo City, Guizhou Province, 27 April.
Lin, Justin Yifu, Fang Cai, and Zhou Li. 1994. "Why China's Economic Reforms Have Been Successful: Implications for Other Reforming Economies." Beijing University, Beijing; Duke University, Durham, North Carolina; Australian National University; and the Chinese Academy of Social Sciences. Draft.
Liu, Gordon, X. Liu, and Q. Meng. 1994. "Privatization of the Medical Market in Socialist China: A Historical Approach." *Health Policy* 27: 157–74.
Liu, Xingzhu, and William C. Hsiao. 1995. "The Cost Escalation of Social Health Insurance Plans in China, and Its Implication for Public Policy." *Social Science and Medicine* 41(8): 1095–1101.
Liu, Yuanli. 1995. "Inefficient Use of Limited Resources in China's Poverty Areas: Problems and Policy Implications." Paper presented at International Seminar on Financing and Organization of Health Care for Poor Rural Population in China, Fragrant Hills Hotel, Beijing, 8–10 October.
Liu, Yuanli, William C. Hsiao, Qing Li, Xingzhu Liu, and Minghui Ren. 1995. "Transformation of China's Rural Health Care Financing." *Social Science and Medicine* 41(8): 1085–93.
Liu, Yuanli, Shanlian Hu, Fu Wei, and William C. Hsiao. 1996. "Is Community Financing Necessary and Feasible for Rural China?" Harvard University School of Public Health, Cambridge, Mass.
Lok, Sang Ho. 1995. "Market Reforms and China's Health Care System." *Social Science and Medicine* 41(8): 1065–72.
Luo, Wujin. 1995. "The Health Financing Study in Rural Areas of China." Paper presented at health care seminar, Beijing, 2–5 May.
Luo, Wujin, Leiya Han, Wei Fu, Ming Wu, Yunjie Song, and Jialing Jiang (China Network of Training and Research in Health Economics). 1995. "Study of Health Financing and Organization in Poor Rural Areas of China." Paper presented at the International Health Policy Program Research, Writing, and Dissemination Workshop, Washington, D.C., 1–9 March.
Luo, Wujin, and Shanlian Hu. 1995. "Study of Health Care Financing and Organization in Impoverished Rural Areas of China." Paper presented at International Health Policy Program Sixth Meeting of Participants, Beijing, 2–7 October.
Mao, Zhenzhong. 1995. "Studies in Sichuan Rural Health Insurance Experiments." Paper presented at health care seminar, Beijing, 2–5 May. West China Medical University, Chengdu, Sichuan.
McEwan, Peter J.M., ed. 1995. *Social Science and Medicine* (Special issue: Aspects of the Medical Care System in the People's Republic of China) 41(8): 1041–1201.
Meng, Jiachi. 1995. "A Report Outline to the World Bank Research and Investigating Group to Raise Funds of the Chinese Health Department." Paper and personal communications presented to World Bank mission in Qiannan Prefecture, 24–29 April. Qiannan Health Bureau.
Meng, Jianguo. 1994. "Pilot Reform of the Separation Between State-Owned Large Enterprises and Their Hospitals." Paper presented at health care seminar, Beijing, 2–5 May. China, Ministry of Health, Beijing.
———. 1995. "Financial Analysis of Medical Facilities." Paper presented at health care seminar, Beijing, 2–5 May. China, Ministry of Health, Beijing.
Murray, Christopher L. 1994. "Quantifying the Burden of Disease: The Technical Basis for Disability-Adjusted Life Years." *Bulletin of the World Health Organization* 72(3): 429–45.
Murray, Christopher L., and Alan D. Lopez. 1994a. "Global and Regional Cause-of-Death Patterns in 1990." *Bulletin of the World Health Organization* 72(3): 447–80.
———. 1994b. "Quantifying Disability: Data, Methods and Results." *Bulletin of the World Health Organization* 72(3): 481–94.
Murray, Christopher L., and Alan D. Lopez, eds. 1994c. *Global Comparative Assessments in the Health Sector: Disease Burden, Expenditures and Intervention Packages*. Geneva: World Health Organization.

———. 1996. *The Global Burden of Disease and Injury Series.* Vol 1. Harvard School of Public Health. Cambridge, Mass: Harvard University Press.

Murray, Chistopher L., Alan D. Lopez, and Dean T. Jamison. 1994. "The Global Burden of Disease in 1990: Summary Results, Sensitivity Analysis and Future Directions." *Bulletin of the World Health Organization* 72(3): 495–590.

Musgrove, Philip. 1996. *Public and Private Roles in Health: Theory and Expenditure Patterns.* World Bank Discussion Paper 339. Washington, D.C.

Novotny, Thomas E., and Richard Peto. 1988. "Estimates of Future Adverse Health Effects of Smoking in China." *Public Health Reports* 103(5): 552–53.

OECD (Organization for Economic Cooperation and Development). 1994. *The Reform of Health Care: A Comparative Analysis of Seven OECD Countries.* Health Policy Study 2. Paris.

———. 1995. *OECD Health Data: A Software Package for the Cross-National Comparison of Health Systems.* Paris.

Palacios, Robert J. 1995. "Reforming Enterprise-Based Social Services in China." World Bank, Washington, D.C.

Parker, R.L. n.d. "Comments on Maternal and Child Health Needs and Strategies for the 1990s in China." UNICEF, New York.

Perkins, Dwight. 1994. "Completing China's Move to the Market." *Journal of Economic Perspectives* 8(2): 23–46.

Peto, Richard. 1986. "Tobacco: UK and China." Letter to the editor. *Lancet* 2(8514): 1038.

Peto, Richard, Alan Lopez, Jillian Boreham, Michael Thun, and Clark Heath Jr. 1994. *Mortality from Smoking in Developed Countries, 1950–2000.* Oxford: Oxford University Press.

Piazza, Alan. 1986. *Food Consumption and Nutritional Status in the PRC.* Boulder, Colo.: Westview.

Picard, John, Lechun Xie, and Qingsi Zheng. 1992. "Cost-Effectiveness of Managing Acute Childhood Diarrhea in Yunnan Province, People's Republic of China." *WHO Research Reports,* pp. 419–24.

Popkin, Barry M., Keyou Ge, Fengying Zhai, X. Guo, Haijiang Ma, and N. Zohoori. 1993. "The Nutrition Transition in China: A Cross-Sectional Analysis." *European Journal of Clinical Nutrition* 47: 333–45.

Prescott, Nicholas and Dean Jamison. 1983. "Health Sector Finance and Expenditures in China." Population, Health, and Nutrition Technical Notes, GEN 14. World Bank, Population, Health, and Nutrition Department, Washington, D.C.

———. 1984. "Health Sector Finance in China." *World Health Statistics Quarterly* 37(4): 387–402.

———. 1985. "The Distribution and Impact of Health Resource Availability in China." *International Journal of Health Planning and Management* 1: 45–56.

Qiannan People's Hospital. 1995. "Report on Financial Situation of Qiannan People's Hospital for the Delegation of the World Bank, Public Health and Finance Officials." Paper presented by Qiu Jinquan to World Bank mission, 25 April.

Qiannan Prefecture, Health Bureau, Maternal and Child Health Center. 1995. "A Brief Introduction to the Health Care Center for Women and Children in Qiannan Prefecture in Guizhou." Danyun City, Guizhou, 26 April.

Qiannan Prefecture, Health Bureau, Sanitation and Anti-Epidemic Station. 1995. "A Report on the Current Situation, Problems, Difficulties with Sanitation and Epidemic Prevention of the Qiannan Prefecture." Danyun City, Guizhou, 25 April.

Ravallion, Martin, and Shaohua Chen. 1994. "An Update on Rural Poverty in China." World Bank, Policy Research Department, Washington, D.C.

Ren, X.S. 1994. "Infant and Child Survival in Shaanxi, China." *Social Science Medicine* 38(4): 609–20.

Schieber, George J., Pierre Poullier, and Leslie M. Greenwald. 1994. "Health System Performance in OECD Countries: 1980–1992." *Health Affairs* 13(4): 100–12.

Shanghai Medical University, Faculty of Health Sciences, and University of Central Queensland, Department of Health Services Management. 1993. "Multidimensional Comparisons and Development Strategies of Chinese Health Care." Research proposal. Shanghai, and Rockhampton, Australia.

Shanghai Medical University, School of Public Health. 1994a. "Financing Rural Health Services in China: Promotion and Evaluation of Innovative Schemes with Special Reference to Poor Areas and Households for the International Health Policy Program." Shanghai.

———. 1994b. Series of Preliminary Studies of Three Poor Counties (Donglan, Shibing, and Xunyi). Team led by Gu Xingyuan. Department of Health Statistics and Social Medicine, Shanghai.

Shi, L. 1993. "Health Care in China: A Rural-Urban Comparison after the Socioeconomic Reforms." *Bulletin of the World Health Organization* 71(6): 723–36.

Sidel, V.W. 1993. "New Lessons from China: Equity and Economics in Rural Health Care." *American Journal of Public Health* 83(12): 1665–66.

Sine, Jeffrey J. 1994. "Demand for Episodes of Care in the China Health Insurance Experiment." Ph.D. diss. Rand Graduate School, Santa Monica, California.

Sun, Aimin. 1995. "Rural Health in China." Paper presented at International Health Policy Program Sixth Meeting of Participants, Beijing, 2–7 October.

Sung, Hai-Yen, Teh-Wei Hu, and Theodore E. Keeler. 1994. "Cigarette Taxation and Demand: An Empirical Model." *Contemporary Economic Policy* 12: 91–100.

Tang, Shenglan, Gerald Bloom, Yueshan Feng, Henry Lucas, Xingyuan Gu, and Malcolm Segall. 1994. "Financing Health Services in China: Adapting to Economic Reform." Research Report 26. Institute of Development Studies, Brighton, Sussex, England.

UNICEF. 1989. *The State of the World's Children 1989.* Oxford: Oxford University Press.

———. 1992. "Children and Women of China: A UNICEF Situation Analysis." New York. Draft.

———. 1995. "Children and Women of China: A UNICEF Situation Analysis." New York. Draft.

United Nations. 1995. *World Urbanization Prospects: The 1994 Revision.* New York.

Wang, Shusheng, Xuesheng Li, Jun Yu, Hui Yang, Tuohong Zhang, Licheng Zhang, Shuo Zhang, Yide Yue, Guy Carrin, and Aviva Ron. 1996. "The Reform of the Rural Cooperative Medical System in the People's Republic of China." Geneva: World Health Organization.

Wei, Ying. 1995. "An Introduction to Health Financing Patterns in China." Paper presented at health care seminar, Beijing, 2–5 May. National Health Economics Institute, Beijing.

———. 1996. "Memo on Revised National Accounts for 1993." World Bank, Human Development Department, Washington, D.C.

Wei, Yu, and Minghui Ren. 1995. "The Important Issue of Enterprise Reforms: Health Care Insurance System."

Gerontology Center of Boston University, Boston, Mass., and China, Ministry of Health, Beijing.
World Bank. 1984. *China: The Health Sector.* A World Bank Country Study. Washington, D.C.
———. 1990a. *China: Between Plan and Market.* Washington, D.C.
———. 1990b. "China: Long-Term Issues and Options in the Health Transition." 2 volumes. Report 7965-CHA. Washington, D.C.
———. 1990c. "China: Reforming Social Security in a Socialist Economy." Report 8074-CHA. Washington, D.C.
———. 1991. *Lessons of Tax Reform.* Washington, D.C.
———. 1992a. *China: Long-Term Issues and Options in the Health Transition.* World Bank Country Study. Washington, D.C.
———. 1992b. *China: Strategies for Reducing Poverty in the 1990s.* A World Bank Country Study. Washington, D.C.
———. 1993a. "China Rural Health Workers Development Project." Staff Appraisal Report. Report 11404-CHA. Washington, D.C.
———. 1993b. *World Development Report 1993: Investing in Health.* New York: Oxford University Press.
———. 1994a. "China: GDP Per Capita." Report 13580-CHA. Washington, D.C.
———. 1994b. *World Development Report 1994: Infrastructure for Development.* New York: Oxford University Press.
———. 1995a. "China Disease Prevention Project." Staff Appraisal Report. Report 14912-CHA. Washington, D.C.
———. 1995b. "China Fiscal Technical Assistance Project." Technical Annex to the Memorandum of the President. Report P-6552-CHA. Washington, D.C.
———. 1995c. "Country Assistance Strategy of the World Bank Group for the People's Republic of China." Memorandum of the President. Report 14454-CHA. Washington, D.C.
———. 1995d. *Investing in People.* Washington, D.C.
———. 1995e. *World Development Report 1995: Workers in an Integrating World.* New York: Oxford University Press.
———. 1996a. *China: Pension System Reform.* Washington, D.C.
———. 1996b. *The Chinese Economy: Fighting Inflation, Deepening Reforms.* Washington, D.C.
———. 1996c. *Poverty Reduction and the World Bank: Progress and Challenges in the 1990s.* Washington, D.C.
World Health Organization. 1996. *Investing in Health Research and Development.* Ad Hoc Committee on Health Research Relating to Future Intervention Options. Geneva.
Xu, Wenbo. 1995. "Flourishing Health Work in China." *Social Science and Medicine* 41(8): 1043–45.
Yang, H. 1994. "More Individual Management of Funds Anticipated: Insurance System Targets Pension Reform." *China Daily,* 27 October.
Yang, Hui. 1995. "Political Research on Reestablishing and Improving the Chinese Rural Cooperative Medical Care System." Paper presented at health care seminar, Beijing, 2–5 May. Beijing Medical University, Beijing.
Yaung, C. 1989. "Current System of Health Care Financing and Delivery in R.O.C. and Its Challenge for Future Development." Paper presented at International Symposium on Health Care Systems, Taipei, Taiwan, China.
Yip, Winnie. 1995. "The Role of Government in Health Care Financing in China." Consultant Report to the China Health Care Financing Mission, 10 April–6 May. World Bank, Human Development Department, Washington, D.C.
———. 1996. "Notes on the Zhenjiang Demonstration Project." Harvard University, Cambridge, Mass.
Young, Mary Eming. 1983. "A Study of Barefoot Doctors' Activities in China." Ph.D. diss., Johns Hopkins University, Baltimore, Md.
———. 1989. "Impact of the Rural Reform on Financing Rural Health Services in China." *Health Policy* 11: 27–42.
Yu, Dezhi. 1992. "Changes in Health Care Financing and Health Status: The Case of China in the 1980s." *Innocenti Occasional Papers Economic Policy Series* 34(December).
Yu, Dezhi, and Xinming Liu. 1995. "China's Economic Reform in the 1980s and Its Impact on Health Care Services." Paper presented at health care seminar, Beijing, 2–5 May. China, Ministry of Health, Beijing.
Yuen, P.P. 1992. "Private Medicine in Socialist China: A Survey of the Private Medical Market in Guangzhou." *International Journal of Health Planning and Management* 7: 211–21.
Zhan, Shaokang. 1995a. "Drug Use and Drug Expenditure." Draft circulated at International Health Policy Program Sixth Meeting of Participants, Beijing, 2–7 October.
———. 1995b. "Financing of Health Services in Poor Rural China." Paper presented at International Health Policy Program Sixth Meeting of Participants, Beijing, 2–7 October.
Zhao, Yuxin. 1995. "Policy Issues on Governmental Funding to Health Work in the People's Republic of China." Paper presented at health care seminar, Beijing, 2–5 May. Health Economics Institute, Beijing.
Zhao, Zhuyan, and Lusheng Wang. 1995. "Present Ways to Raise Funds for Health Care in the Countryside of China." Paper presented at health care seminar, Beijing, 2–5 May. Health Economics Institute, Beijing.
Zheng, Xiang, and Sheila Hillier. 1995. "The Reforms of the Chinese Health Care System: County Level Changes, the Jiangxi Study." *Social Science and Medicine* 41(8): 1057–64.
Zhenjiang, City of. 1995. "Progress Report on the Pilot Health Insurance Reform." Jiangsu Province.
Zhou, Y.S., and others. 1989. "Analysis on the Infant Mortality Rate in China." *Journal of Chinese Demographic Sciences* 3: 35–46.
Zhu, Naisu, Zhihua Ling, Jie Shen, J.M. Lane, and Shanlian Hu. 1989. "Factors Associated with the Decline of the Cooperative Medical System and Barefoot Doctors in Rural China." *Bulletin of the World Health Organization* 67(4): 431–41.
Zimring, F., and W. Nelson. 1995. "Cigarette Taxes as Cigarette Policy." *Tobacco Control* 4(supplement 1): 525–33.